D0772281

NOISE POLLUTION
Impact and Countermeasures

ANTONY MILNE

David & Charles
Newton Abbot London North Pomfret (Vt)

British Library Cataloguing in Publication Data

Milne, Antony
Noise pollution.
1. Noise pollution
I. Title
614.7'8 TD892

ISBN 0–7153–7701–9

Photoset and printed in Great Britain
by Redwood Burn Limited Trowbridge and Esher
for David & Charles (Publishers) Limited
Brunel House Newton Abbot Devon

Published in the United States of America
by David & Charles Inc
North Pomfret Vermont 05053 USA

CONTENTS

PREFACE

This book started life as a Fellowship thesis on industrial noise problems on behalf of NATO's Committee on the Challenges of Modern Society. The present expanded study is the result of several requests to provide a more comprehensive work on the social impact of noise for a general readership. I have therefore aimed systematically to cover all aspects of noise pollution from its many sources in society to acoustics, hearing loss, the physical control of noise, and the constitutional and legal aspects. The book presents the problem as it appears in the late seventies, and takes into account the changes in the law brought about by the 1974 Control of Pollution Act, and the anticipated legislation concerning noise in the workplace.

However, I feel that I ought to make it clear that this work, although it turns a jaundiced eye on certain aspects of urban life, is not intended to be an oblique criticism of capitalism or of industrialisation. It certainly should not be read as a denunciation of economic growth: I believe it is not growth *per se*, but the thoughtless squandering of irreplaceable natural resources that causes problems. The word 'growth' is neutral and can refer to the development of sophisticated new technologies and processes that conserve, rather than exhaust, resources.

Indeed, the curious thing about pollution problems is that they can be successfully tackled only when there is a surplus of wealth in a country. In a sense it is a circular problem; prosperity generates both pollution and the finances with which to tackle pollution. Expenditure in the fight against pollution acts as both a drain on the GNP and a stimulus to it because of the multiplier effect arising from the extra investment and employment in anti-pollution industries. Many people are worried about the cost of acoustic insulation and noise-suppressing devices. These represent a drain on our resources at present and, no doubt, will continue to do so as long as the government imposes strict noise limits. But, in the long run, we ought to break even when the acoustic appliance industry gets bigger, and when monetary losses are minimised by the careful designing-out of noise by engineers and architects at the drawing-board stage.

The tone of this book is both pessimistic and optimistic. There is an encouraging chapter that spells out how it is possible to quieten down a lot of noisy mechanical processes. But possibilities, like exhortations and recommendations, don't necessarily translate into reality. And there is a vast gulf between the passing of a law and the subsequent enforcement of it. A reader of technical journals might feel, with mounting excitement, that the noise problem is close to being solved. Apparently acoustics can be combined with electronics to make 'anti-sound'; artificial reactions to noise can be creatively used in the development of new types of partitions, and so forth. Unfortunately, like peace and plenty for all, the solutions always appear to be around the next corner, and never on the street where we live.

However, it is not beyond the realms of finance and technology to curb the most significant kind of noise pollution – the inconsiderate neighbour or party-goer. The noise made by people, shouting in the streets, revving up cars, etc., is the worst kind because it is unnecessary and yet courtesy prevents one from complaining when one is disturbed by it. Late-night revellers, take heed.

I would like to acknowledge my debt to Mr Harvey and Mr Dove of the Health and Safety Executive for their help and advice. I am also grateful to certain members of the staff of the Institute of Sound and Vibration Research, in particular Professor J. B. Large, Mr W. I. Acton, and Mr D. K. Jones for the use of facilities and periodicals.

I must also thank Professor M. N. Ozdas and NATO's Committee on the Challenges of Modern Society for providing me with the financial backing that enabled me to do the necessary research for the bulk of the work, and the Department of Health for granting me sabbatical leave.

AM

Part One
THE DISBENEFITS OF NOISE

1

THE GROWTH OF NOISE POLLUTION

In November 1977, an international symposium held in London attracted some six hundred learned members of European acoustical societies, members of the WHO, government ministers and research scientists. They gathered to focus national attention on a too frequently neglected type of urban pollution: noise. The message they collectively delivered was one that could not be lightly glossed over: our modern industrialised societies are getting noisier, and if they continue in this way many people will suffer serious psychological and physical harm.[1]

The dangers of noise pollution are no longer a matter of concern only to the select few. In a government-sponsored enquiry made in London in 1948 half of those interviewed made no reference to the subject of noise. Yet in 1962 a follow-up survey showed that as many as 90% were conscious of noise in their environment.[2] The UK Royal Commission on pollution in the late 'sixties estimated that some ten million people were exposed to 'unacceptable' levels of noise, and that this figure could rise to thirty million by 1980.[3] Surveys and studies like these have highlighted an important correlation that has not gone unnoticed by acoustics engineers and urban planners: rising prosperity enables people to become mobile and motorised; it encourages the growth of air traffic; it stimulates the noisy construction industry; it promotes industry's investment in yet faster and noisier machines; it facilitates the expansion of fleets of police cars, ambulances and fire engines – all of which are fitted out with the latest penetrating sirens; and it swells the ranks of ice-cream vans and their attendant musical chimes.

The correlation, unfortunately, does not work in reverse. Momentary falls in the level of prosperity will not necessarily abate the noise problem, which is an inevitable part of western society, and can only be minimised by certain engineering and planning measures (see Part 3). In the meantime we must try to come to terms with the ubiquitous nature of noise pollution, as it is to be found everywhere that human society exists, and not simply in urban areas where it reaches epidemic proportions. Far more people are subject to its pervasiveness than they are to, say, air and water pollution. It is not as easy to control as other forms of pollution, and evidence shows that it is in many cases more costly to abate. On the other hand, it is far more of a psychic irritant than other pollutants. Noise, of course, has no direct physiological effects – it merely annoys the hearer by sending unwanted energy impulses along his auditory nerves. It is rarely as potentially harmful as the more tangible atmospheric contaminants, and if it were to be abated completely overnight it would not, as Bugliarello *et al.* point out, leave any accumulation in the body's internal organs.[4]

But, in spite of its intangibility, noise has many harmful qualities. The first noticeably disturbing effect is the interruption to sleep caused by the sudden or intermittent noise of aircraft flying overhead or cars in the street. Available evidence emphasises the medically damaging effects of prolonged sleep deprivation. Yet a person's sleep is unlikely to be interrupted if the sound level indoors is less that 60dB for aircraft and 40dB for car traffic. (See Part 2 for a discussion of acoustic terms.) Furthermore, it does not take long for a person to become used to impulsive and regular noises; sleep need not necessarily be interrupted even if quite loud noises occur at regular intervals of up to 30 seconds.[5]

There is a distinct relationship between lengthy exposure to noise and the incidence of heart disease, cardiovascular dysfunctions, gastrointestinal disorders and problems associated with endocrine and metabolic functions.[6] Noise can have an effect on the vegetative nervous system and hence on skin conductivity. It can affect the blood circulation, the contraction rate of the pupils, and the secretion of sweat and saliva. Extremely loud and unexpected noise can cause a 'startle reaction', a tightening of the blood vessels that reduces the blood flow to various parts of the body. Acousticians, subject to the most extreme noises (about 130dB) in the line of duty, report some very peculiar sensations. They are unable to walk properly as their sense of balance goes awry and their feet and legs become numb. Others actually feel as if the skin on their chest is flapping about as if in a breeze.[7]

In addition, as we shall see later, many workers in heavy industries seriously risk becoming either totally deaf or at least permanently hard of hearing. Others may become accident-prone and quarrelsome, or they may suffer from migraine and fatigue. Hand-held tools can give rise to 'dead-hand', an ailment that causes pain and numbness in the fingers. There is also occasional damage to the bones of the hand, and finger joints swell up and become stiff.[8] Loud industrial noise can create resonances in the bones of the head and face, making it difficult to think clearly.

All these are sounds that a person can *hear*, and his awareness of them will allow him to take evasive action when possible. But we should not forget that there is an additional range of infrasounds. *Infra* and *ultra* sounds are at the extreme lower and upper ends of the audible acoustic spectrum, and correspond with infrared and ultraviolet light in the visible spectrum. These arise from, for example, powerful internal combustion engines; we are never aware of them because the frequency range of such sounds is beyond the range of our hearing. Infrasounds are extremely insidious, and work on the body's internal organs to set up a vibration which can cause giddiness and nausea at the first stages. When the intensity of the infrasound emissions increases, it leads to a quick but painful death.

It is the cumulative *total* of noise sources – and the greatly increased statistical chance of ordinary people being exposed to some or all of them in their lifetimes – that is so harmful. The consequences of hearing loss are recognised to be more handicapping today than before because the communications media and the telephone play a more important rôle in our daily lives.

We ought, at this stage, to place the hazardous aspects of noise into their proper perspective. The ill-effects of noise described above result from *prolonged exposure*. Noise can cause loss of hearing when it exceeds 90dBa for many hours at a time, but at its simplest level this can be controlled by the use of earmuffs. There are inherent dangers in extrapolating existing urban noise-levels into the future, as the acoustician Dr Vern Knudsen did in 1972 when he said the perceived 20dB increase of the past twenty years would become lethal in another twenty.[9] It is also as well to remember that for as long as man has been blessed with eyes and ears he has been subject to both 'eyesores' and 'ear-sores', and he has had to suffer much more in the past from unpleasant smells arising from inadequate or nonexistent sanitary arrangements. All of these have been accepted as an inevitable feature of daily life.

True technologically-based noise pollution started many thousands of years ago with the invention of the wheel. Indeed, so annoying did the repetitive thud of horse hooves and the rattle of speeding wheels become that Julius Caesar once tried to have daytime chariot racing banned. City life in medieval times was also plagued with the clatter of shod hooves and metal-rimmed cartwheels on uneven cobbled streets. The car is remarkably quiet considering its capacity, power and speed; the real problems lie in the growth in the volume of traffic and the number of traffic-carrying roads.

If noise is an historical feature of civilisation, it must be associated with some of the more pleasing aspects of living within urban communities. In parts of Paris and Amsterdam a certain amount of ambient noise in the evening is considered desirable. Music has throughout history contributed to human happiness, and in the past had much more of a public presence, in the form of street carnivals, than today. Music can be stimulating: it can possess harmonal arrangements, highlighted by strong rhythms and beats, which can be more pleasing to the ear when the loudness is increased. Percussive brass bands, with their thumping bass drums and cacophonous range of high- and low-frequency brass instruments, can offer their musicians a great deal of personal creative satisfaction and excitement.

The term 'loudness' introduces us to the subjective features of acoustic energy – where loudness is often semantically confused with noise.

In view of what has just been said about the pleasant aspects of some noise, community-generated noise is often regarded ambivalently. Unfortunately, the case against noise is much stronger than the one in favour of it. People, for example, sometimes make a noise when they enjoy themselves, but not everyone in their vicinity wishes to participate in conviviality at the same time. This has been as true in the distant past as it is at present. Today people are just as considerate (or inconsiderate) as no doubt they were in the seventeenth century, but the progress of technology has magnified the problem. Sophisticated electronic methods of amplification have led to the rise in complaints about noise from public houses and clubs.[10] Pubs in the UK are licensed to have live or recorded music, but the licensing laws cannot do much about the rowdy behaviour of patrons. Much unpleasant noise is caused directly by people themselves. It appears to most people that of all the varied sources of noise in society the sound of loudly vibrating vocal cords is the one that is most amenable to spontaneous and utterly effective control. As the Noise Advisory Council (NAC) remind us:

> People leaving clubs or public houses in residential areas often seem totally oblivious to the fact that there may be children or adults sleeping close by. Enlivened no doubt by the intoxicating liquor they have consumed, they shout loudly to their friends, often for long periods after the premises themselves have closed. Car doors are banged and car engines raced in the most unnecessary manner. We return again to the question of good behaviour generally; this is where a little consideration and thought can prevent considerable disturbance to others.[11]

So, despite the growth of technologically-generated noise, a great deal of disturbance is created by people rather than by machines. A surprising amount of noise is made by individuals who seem not to be conscious of the nuisance they are causing, and who could avoid creating the noise at very little expense to themselves. For example, lorry drivers could remedy brakes that squeal, loose loads that crash against the cab when braking, rattling bodywork and loud horns. Those bent on drawing attention to themselves by revving car and motorbike engines and over-using horns might be more considerate. Noise used as a form of advertising can also be an irritant: both boutiques and discotheques amplify music into the street to attract (and in some cases repel) passers-by; combined with the very dense traffic flow, the rise in the general level of noise can considerably detract from the pleasure of shopping in such areas.

However, it is when technological and man-made noise comes together within the home that quite ordinary domestic activities sound much worse than they used to only decades ago. Gardening enthusiasts can now use cheaply made electric lawn mowers and D.I.Y. experts can buy a range of electric drills as well as the usual panoply of grating and knocking tools. As Tony Aldous writes, 'Convenience and cheapness to one man is bought at the expense of a neighbour's tranquillity . . . . It is the aggregation of trivial noises which goes to make up what is often a gross affront to the environment . . .'[12]

Today, the growing problem of noise pollution is no longer one purely of scientific interest to acoustic physicists. The British government itself considers 'unnecessary and intrusive' noise to be a form of pollution,[13] and therefore authorises the payment of noise-insulation grants to one-tenth of British homes.[14] But unlike other forms of pollution, the growth of which depends on somewhat specific industrial processes, pollution through noise is certain to increase because it arises not merely from industrial development but also from urbanisation. As urban communities develop so air and road traffic movement increases.

In the immediate past, when air contamination in built-up, industrial areas was a serious threat to health (the excesses of which have now been considerably abated), there were few cars, and even fewer airports. Noise is hence much more a function of population growth, regardless of its specific source. In other words as populations expand and urban areas spread outwards, and as both industry (to a gradually lesser extent) and road and air traffic (to a greater extent) increase either in size or volume, the sources of noise grow commensurately. (It is as well to remember that as Britain's industrial base is in a state of gradual decline, purely industrial pollutants will also diminish.) In the USA, for example, the worst of the urban noise problems affect 130 million people now living in metropolitan areas, and large communities around major airports like the heavily used O'Hare Field in Chicago, Kennedy Airport, New York and the international Airport at Los Angeles are most seriously impacted.[15]

Noise pollution is a complex social phenomenon. Noise from industry and noise from the community itself are undeniably linked as part of the urban problem. There appears to be no part of man's working or leisure periods when he is free of the impact of noise. The assembly-line worker who is subject to about 90dB at his job still has to experience about 50–60dB when he emerges from his place of work into the general environment.

The determination of excessive noise levels is almost beyond the scope of scientists. Noise is a human and environmental phenomenon and needs the sociologist and the planner to help to isolate some of its intricate and subjective features. The degree of annoyance caused by noise has been measured in both British and US surveys.

It has been suggested that individuals react subjectively to sound waves according to whether they are more introvert or extrovert. Some people may seek more aural stimulation than others. Noisy (especially musical) environments may be quite enjoyable to them for longer periods.[16] Loudness (as opposed to sound intensity) is a subjective experience, but there are now methods of measuring sensations of both loudness and annoyance. In 1972 a research report, after reviewing tolerance levels internationally sought to establish a common ground between 'tolerable' and 'intolerable' noise levels. It found that no such common ground could be devised because the ranges between the two extremes were so wide. Such knowledge serves to illustrate to the planner the difficulties in deciding on reasonable degrees of noise people should put up with.[17]

The problems of urban societies have been noted by writers like Elaine Morgan who are concerned about the environment. It is the complex interdependency of core features of city life that seems to act as a catalyst for any incipient hazards. 'Anything that happens,' she writes, 'from the construction of a new airport, or the decline of the cinema . . . doesn't only affect the lives of the people immediately concerned but starts a whole chain of reaction of economic side effects.'[18] In regard to noise

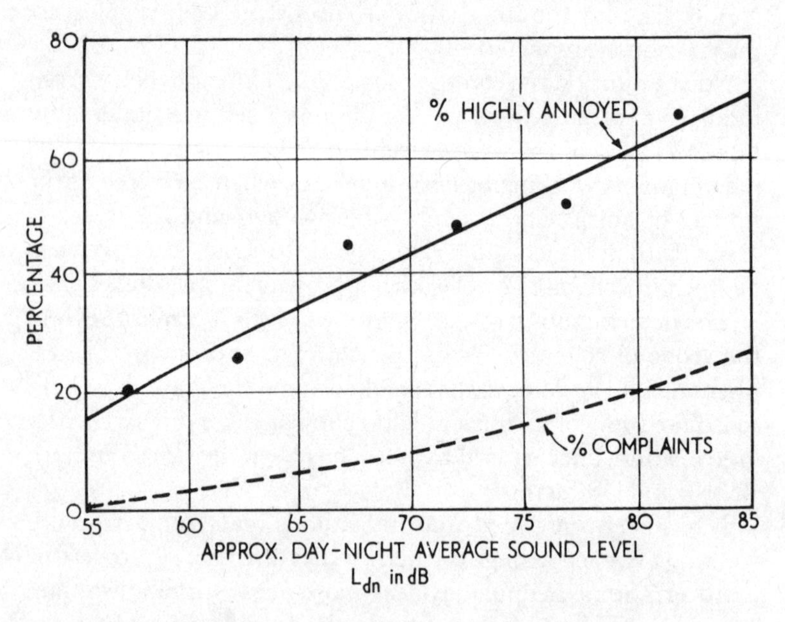

Figure 1 Percentage of population reported to be highly annoyed as a function of day–night average sound level. Combined results – British and US – surveys. Source: Noise–Con 75 Proceedings, US. Nat. Bureau of Standards, 1975

pollution the disamenity aspect becomes less one of complex interdependency than of direct physical interrelationship. The density of some suburban localities causes noise from buildings to spill out into the streets. In the absence of proper containment policies the easy solution is simply to ban the activity complained of, and thus perhaps deprive many individuals of innocent pleasure.

Unfortunately technological progress means that the catchment area into which noise is disseminated gets wider. Amplifiers are used at pop festivals and other open air events, often until early in the morning. Many private parties now have amplified equipment which can be hired from mobile discotheques. This sort of music in private households can easily reach 80dB, which is considerable in view of the confined spaces involved. Most pop groups operate at well over 90dB, and at one pop festival a sound level of 116dBa at a distance of 98.4m (300ft) was recorded.[19] Although this is an extremely high level of noise, it is likely to cause hearing damage only to those in the closest vicinity. According to the NAC the dangers of harm from this sort of noise may have been exaggerated. They admit that much of the research on hearing loss from loud music is inconclusive, and urge more study of the subject.[20] But there is far more empirical evidence about the harm caused to professional musicians. For example, Roger Daltrey, lead singer of the Who, in considering whether to finally split up the group, said, 'Right now whether we ever tour again is up to Pete (Townshend). The doctors have told him that if he keeps on playing he will be deaf by the time he's 40. It's all down to the years and years of loud music. I mean, even my hearing has been damaged. I've lost the whole top register of my hearing, which means I just can't hear high notes.'[21]

Public concern about noise levels has possibly now reached the stage where the definition of a sound as 'noise' automatically makes it a target for control. But this kind of overreaction ought to be resisted. Hitherto noise itself was generally referred to in a negative sense, so pleasing sounds (like music, or the sound of rustling leaves) had to be isolated and specified.

But noise ought always to be used as a subjective, rather than as a pejorative term. It should be acknowledged that there is a wide range of continuous and background 'noises' (including background music) and that people can find a noise pleasing one day and irritating the next.

The policy-making planner thus has to be cautious in his approach to containing various social disamenities. An anti-social activity may be correlated with noise, but the noise itself may not cause the anti-social activity. The Ice-Cream Alliance, a body representing small ice-cream manufacturers, told a working group of the NAC that it was not the chimes of ice-cream vans that annoyed residents so much as the repeated demands of children for ice-cream.[22] The town council, in refusing planning permission for a new club or disco, has to be sure about whether it is acting in the interests of a new noise abatement policy or in connection with some other less easily definable public nuisance. A pub that plays disco music may attract rowdy and troublesome youths, and be the focus of periodic complaints from neighbours. But the source of the public nuisance is often the clientele itself rather than the amplified music.

The next two chapters concentrate on the specifically mechanical sources of noise that increasingly plague urban areas. But it is hoped that this chapter, by focusing attention on the societal aspects of noise, has been a salutory reminder of the human impetus behind many of the irritating, and indeed disturbing, noises to be heard virtually around every street corner.

2

THE PRICE OF MOBILITY

According to enquiries into subjective reactions from people in many countries, the noise of road traffic is a major problem that the town dweller now has to face. In some places, notably in the UK, France, Scandinavia and Japan (all with highly congested urban areas), traffic noise can be *the* major source of public annoyance. Significantly, these countries all have high standards of living; car ownership for most families is nowadays common. As Tony Aldous points out, motoring for the Everyman tends to be a 'fifth freedom', and is as entrenched a part of an Englishman's civil rights as the more traditional freedoms.[1] At least in theory it facilitates total freedom of movement and independence from public transport and all its deficiencies – the main constraint is purely financial. In many regions of the USA distances between food shops, schools and work are so great that a car is a virtual necessity.

According to a government social survey carried out in London, the sound of road traffic is unquestionably the most important *source* of noise bothering people both at home and out of doors. Indeed, 36% of people at home and 20% of those outdoors were irritated by it.[2] A report of the working group researching traffic noise for the NAC estimated in 1970 that 46% of the urban population (about 21 million people) lived in dwellings with external levels of 65dB or more, which is above that considered to be suitable for residential districts. Of this total, some 8½ million people were exposed to an external level of 70dB or more. Even in a rural area with a population of 9 million, 32% were said to be afflicted with an external level of

65dB, and a further 19% to a level of 70dB.[3]

Central London naturally has one of the biggest noise problems in Britain, and there beleagured commuters and shoppers are regularly exposed to 70 and 80dB levels exceeded only by motorways and airports.[4] The noise made by a motorcycle can be particularly penetrating in city areas, and can rise as high as 120dB. At a distance of 16.4m (50ft), i.e., well within the vicinity of private homes, current models can yield 77 to 88dBa, and older machines can emit levels as high as 96dBa.[5] Traffic noise is now largely continuous for long periods during the day and well into the night. Tests from the Post Office Tower conducted by the GLC Scientific Branch suggest that the quiet period during the night when traffic is minimal is gradually getting shorter.[6] The statistics about the striking growth in the number of the world's vehicles dramatically highlight the potential for universally louder noise levels. As the London enquiries of 1948 and 1962 illustrated, the growth in the number of people becoming exposed to traffic noise is of almost geometric proportions. As a further example, the size of the US urban population experiencing levels above 55dBa will have increased by a factor of four between 1970 and 1985.[7]

Today there are 300 different car models among the seven million produced annually by 16 big European companies, and there are ten million manufactured in the US by four makers.[8] According to the Organization for European Co-operation and Development (OECD) the number of vehicles in the world rose from 100 million units in 1960 to 200 million in 1970, and these were expected to grow in numbers to reach 300 million by 1980.[9] In France alone, the OECD expect the volume of cars to rise by 55% between 1970 and 1985, which would roughly match the percentage growth in the US during the past ten years. And we should bear in mind that the increase in the numbers of cars on the roads is accompanied by the greater mileages being covered by individual motorists. This factor suggests that the noise levels rise more in suburban areas than in city centres, which often unwittingly inhibit speed and noise levels through heavy congestion.

Inevitably, heavy commercial vehicles cause the greatest annoyance and are almost universally disliked, although they comprise only 5% of the vehicles on British roads. It is generally felt that they contribute more than their fair share to the total amount of noxious fumes in urban streets, as well as to the congestion and insidious vibrating forces that disfigure and slowly destroy the fabric of small provincial town centres with their narrow thoroughfares. The number of lorries in the UK, in fact, has increased fourfold in just eight years, and the maximum payload they may carry has been raised because of the UK's entry into the EEC. Research by the Road Research Laboratory hints that the number and magnitude of heavy axle loads is steadily increasing.[10]

As we will see later, diesel engines are inherently noisier than their petrol counterparts, and many public transport systems in Britain and abroad consist essentially of diesel powered buses. In the USA there are almost 120,000 miles of bus routes, where engine noise-levels rise to 94dB.

Engine noises are at their peak in the high-revving lower gears. The NAC's Panel on Noise in the Seventies is now devoting a lot of attention to the problem of city vehicles accelerating from standstill in busy streets. The Panel believes more attention should be paid to the method of both construction and operation of buses, taxis and lighter commercial vehicles. In view of what is now known about traffic volume and the increase in decibel emissions, the Panel thought it essential to reduce commercial vehicle noise by 5dBa before 1976, otherwise levels would rise inexorably.[11] The government, acting on the advice of the NAC, are intent on achieving at least long–term results in this matter. Permitted emission levels for new vehicles will progressively be lowered, in spite of the extra work in vehicle design this will entail for manufacturers (and no doubt in spite of the extra cost).

As with other aspects of the world's environmental problems, the phenomenon of growing car ownership improves the quality of life in some ways and harms it in others. And yet as a general principle one could argue that the disbenefits are never so widely appreciated as the benefits. One might even agree with E. J. Mishan's assertion that the pervasiveness of the motor car is so closely bound up with our way of life and habits of thought that its intrusion is barely noticeable.[12] But Mishan was writing in the late 'sixties, so how true is this ten years on? There are now pointers to an opposing ethic, where amenity action groups are obliging the authorities to adopt stricter routing and road diversion policies. Indeed so rigorously successful has the GLC been in this sphere that it has now gained itself a reputation as the most punitive anti-motoring authority in Britain: parking and free access to central and inner London areas has been severely curtailed.

Should the authorities expect the numbers of vehicles on British roads to keep on growing until a logical optimal limit is reached where every adult member of the population is in possession of a car? Michael Allaby suggests that none of the political parties would welcome such a situation, and that they all agree that no feasible system of roads could accommodate the consequent mammoth volume of traffic.[13] In the future there will no doubt be more public pressure to prevent the closure of more railway lines, and there will be attempts to improve public transport services. The government as a whole is far more interventionist in matters of social amenity and public health than it used to be.

The powerful motoring lobby and the road haulage industry, with its vested economics interests, are being hard put to maintain the advantage over *both* the environmentalists *and* the authorities. None of this of course will please the motorist, but, if he is fair-minded, he will realise that a balance of equities has to be maintained in the interests of the community and of himself. When the rights of the motorist to totally uninhibited mobility begin to *seriously* infringe the rights of householders to an unspoilt and quiet environment, then certain measures to redress the balance must be taken.

One basic difficulty with the setting of standards to contain excessive traffic noise is that a lack of understanding of the intricacies of noise abatement in the past has meant that permitted levels have been too lenient. For example, in 1966 the GLC tested a number of heavy lorries with an audiometer and not one exceeded the then prescribed limit.[14] There were only two prosecutions in 1970 under the 'exceeding maximum permitted sound level' provisions of the Motor Vehicles (Construction and Use) Regulations of 1969.

It is by no means easy to acquire suitable testing sites which are not surrounded by noise-reflecting buildings and structures. Neither is it possible to get hold of sophisticated electronic measuring equipment or properly qualified personnel in sufficient quantities. In the meantime there is a lack of rigour in enforcing higher construction standards for vehicle silencers which are nowadays made to 'built-in-obsolescence' standards, and disintegrate in only a few years. It is estimated that at any one time there are about 250,000 cars on the road with defective silencers.[15] Faulty silencers are in fact responsible for the bulk of motor noise prosecutions (there were 13,807 in 1970 in the UK alone).

It is regrettable that the government has been rather lax in the matter of setting enforceable standards to contain vehicular noise. Of course, merely limiting the number of vehicles that are actually manufactured or sold in the interests of conservation would be an unwarranted restriction on economic growth and the rights of individuals. But limiting the *volume* of traffic on certain roads or in particular areas is a more acceptable solution. The stringent parking policy of the GLC has much to commend it, and, although it curtails drivers' rights in one sense, it has the potential of speeding the flow of through-traffic. And by obliging commuters to use public transport, it saves them wasted time in traffic jams and excessive engine wear-and-tear.

However, although it is recognised that lorries are the chief offenders, many obvious measures that could have been taken have gone by default. British Rail's unsuccessful Freightliner service, where goods were taken on the remaining road journey after traversing the country by rail, suffered from financial, promotional and planning handicaps. There has been a failure on the part of local authorities to ban lorries from obviously sensitive residential areas. Because of financial cut-backs in the UK, many villagers must continue to put up with the tortuous negotiation of single-lane high streets by heavy Continental and British container lorries that ought to be using major bypass carriageways. Because of shortage of money, the authorities have also tended to neglect the regular resurfacing of roads, and the maintaining of road surfaces in good repair is one way of minimising traffic-induced vibrations at the source.

After years of negotiations, the first regulations on traffic noise finally became law in 1968 when the maximum noise levels for lorries were fixed at 89dB, and 84dB for cars. The Minister of Transport later admitted that these limits were too lenient, but introduced an argument against stricter standards that will recur from time to time in this book; that of economic and engineering feasibility. Nevertheless, in 1974 the limits were reduced to 86dB for lorries and 80dB for cars. Although these figures appear to be negligible, they do in fact represent quite realistic reductions in view of the logarithmic progression of the decibel scale (see Part Two). The question now remains as to how well these limits are enforced in practice.

Aircraft noise is a nuisance second in importance only to road traffic noise, and people generally have become aware of this additional irritant in their lives since the widespread introduction of jet planes in the early 'sixties, which has been aggravated of late by the package-tour explosion. A 1961 government survey of the number of persons annoyed by aircraft flights in a 10-mile catchment area around Heathrow was 378,000 out of 1½ million.[16] Some specialists say that around this particular airport some four million people are affected.[17] Others claim that 10% of the population are disturbed by the noise at Heathrow alone (i.e., 5½ million people).[18] In a subsequent survey in 1967, the government found that over the years more airport zone residents were rating the district where they lived as being subject to a nuisance that was likely to get worse.[19]

Aircraft personnel who work on the tarmac directly under the flight paths of planes coming and going to and from Heathrow Airport are subjected to a deafening level of 115dB. And if it were not for the fact that these employees are obliged to wear ear-muffs, they would all suffer substantial degrees of hearing loss within a few months. It is not surprising, therefore, that the loudest single source of technologically produced noise should give rise to concern about its environmental impact.

Table I: Distribution of Noise-annoyed Airport Residents.
Source: *The Impact of Noise Pollution*, Bugliarello *et al.*, Pergamon Press, 1976

	Percentages			Absolute Numbers	
PNdB Stratum	% of Total Population in Stratum	% of Stratum Annoyed*	% of Total Population Annoyed*	No. of People Annoyed*	No. of People in Stratum
103+	3	68	2	28,000	42,000
100–102	6	51	3	42,000	84,000
97–99	7	48	3	42,000	98,000
94–96	13	36	5	70,000	182,000
91–93	27	24	6	84,000	378,000
88–90	22	23	5	70,000	308,000
85–87	11	16	2	28,000	154,000
Up to 85	11	10	1	14,000	154,000
Total	100		100	1,400,000	1,400,000

* Score on the annoyance scale of 3.5 or above.

There are important and subtle psychological overtones, too, in the problem of aircraft noise since people, as well as resenting the noise, are apprehensive about plane crashes in their area. The volume of air-traffic movement is growing exponentially. Unlike the deliberate planning policy of extending the motorways and thus 'siphoning-off' through-traffic, the growth in the volume and frequency of air flights has far-reaching effects. The hazard becomes more apparent to families that tend to move near airports because they are constrained by housing shortages. In the Heathrow area, the population between 1963 and the early 'seventies increased by around 30%. In addition to take-off areas becoming larger, the general servicing facilities for travellers gets bigger: this entails the need for more catering and ancillary staff, who thus add to the numbers exposed to the noise of the airport. Such staff often live in the airport's catchment area.

Citizens the world over have endeavoured to intervene in the decision over the siting of airports. Both in Britain and in the USA people have petitioned the authorities responsible for plans to site third airports for the major commercial cities of London and New York respectively. In the USA an earnest search is being conducted for suitable sites, and even Long Island sites and swampland in New Jersey have been considered. The site is now expected to be 50 or even 70 miles away from New York City, and the Federal Aviation Commission is investing $390,000 to study the feasibility of an off-shore jetport.[20]

Turbojet engines are noisier than most but they are more efficient at supersonic speeds, so aircraft engineers have unfortunately had to revert to them. However, since 1970, subsonic planes have been required, by the International Civil Aviation Organization (ICAO), to satisfy noise certification requirements before they can enter service.[21] New designs should be only half as noisy weight-for-weight as designs dating from 1969 or earlier; most countries agree with this ruling. Even so, the NAC are not convinced that this agreement will necessarily bring the existing noise levels down. They feel it will be some years before the present generation of turbojet aircraft, such as the VC10 and the Boeing 707, is replaced. Indeed, by 1980 about half of the planes in service around the world will still be of the four-engined large turbojet variety, and the remainder (i.e., those similar to the Boeing 747) will be using the quieter high–bypass turbofan engines.

Technological developments will improve matters in the long-term. The projected new generation of planes, with engines like the RB211 and the revolutionary short and vertical take-off and landing craft (STOL and VTOL), could, if extended to passenger flights, use existing airports, or only part of them. They will still need to overfly residential areas (but within a tighter zone) until they gain their optimum height.

Attempts have been made to reduce aircraft noise through the re-routing of flight paths. In 1971 a working group of the NAC suggested that Minimum Noise Routes (MNRs) should be set up which would concentrate departures along the fewest number of routes possible. This could intensify the problem for those communities unlucky enough to be overflown by the MNRs, but the public as a whole have consented to the idea and are critical mainly of those aircraft that do not adhere strictly enough to the prescribed routes.

Sonic booms are an additional hazard to those living near

Table II: Estimated Aircraft Noise Levels.
Source: 'Noise in the Next Ten Years', Copyright, HMSO, 1974

	Take-off—Flyover[1]			Landing[1]		
Aeroplane Type	Existing	With Retro-A[2]	With Retro-B[2]	Existing	With Retro-A[2]	With Retro-B[2]
VC 10	110	106		115	110	
DC-8	115	103	93	116	104	101
Boeing 707	114	103	93	119	104	101
Boeing 727	99	95	91	109	99	96
Trident I	100	96		108	104	
Trident II	109	105		109	105	
Trident III	105	103		110	104	
DC-9	98	93	90	110	100	98
Boeing 737	101	92	90	111	102	99
BAC 111	98	92		104	98	
[Concorde][3]	114	—	—	115	—	—
[Boeing 747	108	—	—	107	—	—
DC 10	99	—	—	106	—	—
Tristar]	97	—	—	103	—	—

NOTE: [1] Estimated in accordance with ICAO Regulations.
[2] Retrofit A—Moderate noise treatment to nacelle and jet.
Retrofit B—More extensive noise treatment including new front fan modification.
[3] Figures from manufacturers included for comparison.
—Not applicable.

flight paths, and are likely to become a growing nuisance. In the USA they have been a serious disamenity for many years, where SST schedules have exposed between 35 million and 65 million people to between 10 and 20 booms a day. In 1968, Oklahoma was subject to 1253 SS flights during daylight hours over a period of six months. In spite of an opinion poll which found that 73% of residents had become tolerant of the noise, 12,588 complaints were lodged with the authorities, 8335 of them alleging damage to property.[22] Some 6000 people complained bitterly about eleven sonic bangs over southern England by Lightning fighters in the late 'sixties. As Kryter has pointed out, booms from SSTs are no more hazardous than noise from subsonics, but they have a much wider impact[23] as the shock wave of the boom continues to travel over the ground in the wake of the aircraft. In fact this shock wave can cause more harm to sheep (which have been known to abort) and to buildings (where it can crack stucco plaster and break windows). It is significant that the Greek government has banned SST flights over the Acropolis and other classical monuments.[24]

A report from the Institute of Sound and Vibration Research concludes that people are far less concerned about railway noise than they are about any other. There is no official British system of measuring acceptable noise levels from railway lines, but one interesting discovery shows that whenever attempts are made to apply the road and air tests (in the form of standard official measures) to railway-side communities it is found that the objective sound recordings elicit a much lower subjective response than expected.[25] There may be an inevitable element of resignation among such communities: most railway lines have been established for many decades; there is no rapid growth in rail transit paths as there is in aircraft flight paths; and railway noise does not increase commensurately with speed.

Apart from the obvious noise source of the engine and track, trains also make an unpleasant noise because of shunting operations, whistles and hooters. There is some evidence that British trains (at an average level of 70dBa) are certainly noisier than their European counterparts. There has been a considerable increase in the number of railway networks of many kinds, especially in Japan, Europe and the USA, which often include elevated surface transit systems.

Members of the Surrey and Kent Action on Rail (SKAR) feel that insufficient attention has been paid to the problems of noise and track landscaping. They are also concerned about the unknown elements of British Rail's new high speed trains. The Japanese are paying out millions of yen to soundproof the track and the houses along the Shinkansen 'bullet-train' lines, and British Rail are not unaware of the insulation and compensation factors entailed in their own more recent train innovations.

3

INDUSTRIAL NOISE

As we saw in Chapter 1, noise is a function of urbanised and *industrialised* society. So most citizens suffer from emissions from workshops, factories and industrial estates, as well as from traffic and aircraft, at some time or other in their lives. Noise arises, 90% of the time, from either manufactured goods or from the manufacturing process itself.

The persistent irritant of factory noise being emitted into a residential neighbourhood is one that local authorities are tackling as best they can. The situation is worse for those districts that are not naturally shielded, or protected by administrative buildings, which can in fact be erected as a purpose-built screening device by percipient town planners. In spite of the decline in dense, heavy industrial areas with their attendant 'back-to-back' accommodation along narrow terraced streets, more houses are still being occupied by people in some newer areas who have no connection with the nearby factories.

The significance of the industrial noise nuisance is confirmed by the NAC who say that just under half the local authorities in England and Wales have reported 'serious' industrial noise problems.[1] The Public Health Inspectorate have published figures suggesting that half the complaints about noise from commercial and manufacturing premises were confirmed after investigation.

The seriousness of the impact of industrial noise is personally verified by those employees who have to spend much of their working life *captively* exposed to noisy conditions. Working in a noisy environment is the best way of impairing one's hearing.

An early Minister of Housing once pointed out that, if one is planning to control pollution in society, it must be borne in mind that atmospheric contaminants and noxious gases are more of a pervasive hazard to the general public than industrial noise, which is more of a localised, occupational phenomenon.[2] Furthermore, noise pollution at work can result in some alarmingly costly bills. The WHO estimated that, in 1969, the financial losses resulting from accidents, absenteeism, inefficiency and compensation claims attributable to industrial noise in the USA alone amounted to four billion dollars.[3]

Surveys conducted over many years in various countries generally confirm that the problem is serious. Not only are there high percentages of workers involved, but hearing losses in excess of 45 and 50dBs are common. Such degrees of disability mean that the sufferer will *permanently* experience, in normal, non-industrial settings, difficulty with loud conversation at any distance over 1·5m (5ft). Table III shows the gradations of hearing from 15 to more than 85dB and their practical effects.

Exactly how many people are critically affected by noise at their place of work? Expert opinion tends to present a varied and imprecise picture. One difficulty is that in the past hearing loss surveys have not distinguished carefully enough between those suffering from presbycousis (hearing impairment due to aging or medical reasons), and those whose impairment can be directly blamed on the nature of their employment – sociocusis. (The relationship between presbycousis and sociocusis can apparently be clarified by subtracting the average loss expected at a given age from the overall loss at that age, and the net impairment can be averaged with the corrected losses of other similarly exposed people.) Furthermore, there is a somewhat natural tendency for vested economic interests to understate the extent of the problem. For example, only about 25% of British industrial managers think that *some* employees may have cases of occupational deafness.[4] Only three out of fifty-five large firms were going ahead with 'systematic and extensive noise reduction programmes'. Official figures that reflect employers' attitudes to occupational noise must be treated with caution. The US Department of Labor's estimates of $8\frac{1}{2}$ million workers exposed to 85dBa or more [5] may not have accurately stated the position, because its figures were based solely on management sources.

Table III: Hearing loss classification table
Source: 'Noise', Alan Bell, WHO, 1966

Class	Degree	Average dB loss at 500, 1000, 2000 Hz in better ear	Remarks
I	Normal	Less than 15dB	Within normal limits
II	Near normal	15–25dB	No difficulty with ordinary conversation at up to 20ft
III	Mild loss	25–40dB	Difficulty with ordinary conversation when distance exceeds 5ft
IV	Moderate loss	40–65dB	Difficulty with loud conversation when distance exceeds 5ft
V	Severe Loss	65–75dB	Difficulty with shout when distance exceeds 5ft
VI	Profound loss	75–85dB	Difficulty with shout at less than 5ft
VII	Almost total loss	More than 85dB	Loss of practical hearing for speech

In the UK, the Industrial Injuries Advisory Council (IIAC) admitted that they received widely varying evidence on the number of people with noise-impaired hearing, some unsupported, designed rather to illustrate the general effect of a particular working environment than to establish the exact number of people affected. The most recent estimate for US exposure figures has been made by Bolt, Beranek and Newman in a special report prepared for the Occupational Safety and Health Administration (OSHA). The figures were conjectural, but the US authorities accepted their credibility. These figures are given in Table IV.

The British Factory Inspectorate carried out a survey of one

Table IV: Estimate of the number and percentage of US workers over exposed to noise (based largely on informal discussions with spokesmen).

Source: Bolt, Beranek & Newman Inc., 'Impact of Noise Control at the Workplace', Report 2671, p C-2, 1974

Industry	SIC Code	Production workers (thousands)	Overexposed workers (thousands) 85dB	90dB	Percentage overexposed 85dB	90dB
Food	20	1170	820	350	70	30
Tobacco	21	63	48	40	76	63
Textiles	22	900	855	765	95	85
Apparel	23	1174	12	0	1	0
Lumber and wood	24	542	542	390	100	72
Furniture & fixtures	25	427	235	58	55	15
Paper	26	557	395	206	71	37
Printing & publishing	27	661	132	99	20	15
Chemicals	28	596	137	66	23	11
Petrol & coal	29	117	58	23	50	20
Rubber & plastics	30	531	266	106	50	20
Leather	31	256	3	0	1	0
Stone, clay & glass	32	555	416	139	75	25
Primary metals	33	989	577	259	58	26
Primary steel	331	485	325	170	67	35
	332					
Foundries	336	275	189	54	70	20
	333					
Primary nonferrous	334	233	63	35	27	15
	335					
Fabricated metals	34	1123	786	225	70	20
Machinery except electrical	35	1366	956	273	70	20
Electrical machinery	36	1370	959	274	70	20
Transport equipment	37	1354	880	284	65	21
Utilities	49	627	445	188	71	30
Total		14382	8524	3755	59	26

hundred factories which between them employed a total of 16,048 individuals who were subject to the Factories Act (i.e., the survey excluded office workers). Using these sample figures, the Health and Safety Executive (HSE) made some calculations and estimated that out of a total of 6,402,000 employees, some 1,161,000 (i.e., 18·3%) were exposed to 90dB or more for either all or part of the working day.[6]

Table V. Estimated numbers and % of people exposed to noise in manufacturing industry in the United Kingdom.
Source: A. R. Dove, HSE, 1977

	People exposed to noise all of the time		People exposed to noise some of the time[1]	
Sound level (dBA)	%	No. (000s)	%	No. (000s)
less than 80	38.8	2481	5.9	380
80–84	13.9	888	3.5	224
85–89	11.0	704	3.2	203
90–94	6.0	382	3.2	198
95–99	2.7	175	3.5	223
100–109	0.4	27	1.8	114
110 & over	0.1	5	0.6	37
Not measurable[2]	5.6	360	—	—

[1] Less than 6 hours' exposure per day, with measurements based on the most significant noise exposures during the day.
[2] Due to difficulties with sound-meter measurements in certain environments such as those with flammable atmospheres or highly mobile workers.

There is now enough evidence to prove how harmful unchecked industrial noise can be to those who have to earn their living exposed to it daily, for years at a time. With the recent and growing body of knowledge about the deleterious effects of pollution generally, people have become concerned about the disagreeable aspects of industrial activity. In some respects a concern with industrial noise hints of a subtle shift in long established *laissez-faire* attitudes to industry. Indeed, the clatter of machinery, far from being an irritant, sounded sweet to the industrial barons of the nineteenth century. And the output of noise has not grown proportionately with the rapid nineteenth- and twentieth-century spread of methods of production.

The environmental expert often has the unenviable task of drawing attention to a harmful aspect of urban life that hitherto people had hardly noticed. Noise is often tolerated by employees because the industry they have *volunteered* to work for has been noisy in the past, and the conditions in which they find themselves seem to be established as normal, and even inevitable. Furthermore, industrial deafness as such has been known for over two hundred years. In *De Morbis Artificum* (1713), Ramazzini described how those engaged in hammering copper 'have their ears so injured by that perpetual din . . . that workers of this class become hard of hearing and, if they grow old at this work, completely deaf.'[7]

Viewed historically, workers themselves have never attached much importance to the subject, since noise was only peripheral to other disadvantages of working in industry. And the recent evidence about the shortage of skilled labour in Britain shows that it is due to the reluctance of school leavers to take up apprenticeships in one of the skilled trades because of the traditional lack of congeniality in industry as a whole.

Nevertheless, the trade unions and industrial managers may be obliged to accord the issue of noise abatement a higher priority than they have done in the past.[8] There is in fact an abysmal record on the part of all parties concerned in actually controlling noise either voluntarily or compulsorily. The TUC displays indifference by allowing its member unions to make their own agreements with those industries under official pressure to reduce noise levels.[9] Noise control is often brought in as an expensive and ineffective afterthought, so industrial planners are generally content with the status quo. Most by-products can be converted into useful commodities; but the by-products of noise measured in energy terms are non-existent, as we shall see in Part Two. When industry is obliged to work with nil or negative returns it understandably feels reluctant to voluntarily reduce the efficacy of its energy sources.

The industrialist will sometimes (justifiably) point to innovative measures that offer built-in solutions to the noise problem. New machines are introduced that make new noises and meet with greater approval. Noise levels are also noticeably reduced as a routine measure. There is also the automatic replacement of machines with fewer moving or non-metallic components.[10] William Burns has drawn attention to the changeover to welding and the decline of riveting, and to the substitution of metal rollers in assembly lines by nylon or fibre components.[11] In any event, there is a natural decline in the number of workers exposed. It is in this sense that industry gains a human face – through its own advances.

4

THE COSTS OF NOISE

How much does noise pollution cost the country? It is unlikely that anyone will be able to satisfactorily answer this, perhaps because it is rather an inadequately defined question. The reason the question is asked at all is because of the general awareness of some kind of relationship between monetary values and noise.

One big difficulty that has historically confronted economists is the intangibility of social welfare. The impact of social amenities and disamenities often have to be viewed subjectively, because a price cannot be placed on them. If quiet – or clean air – is free, how is its loss to be evaluated? It is disturbing to note that the simple things of life are no longer as ubiquitous as they once were; but this makes the economist's job a little easier. After all, economics is the science of scarcity. Moreover, whereas in the past the improvement in the quality of life was a direct outcome of economic growth, it may not be so today. Indeed, society's welfare may begin to decline as industrial output continues to rise, bringing with it more extensive pollution. On the other hand we should bear in mind that the indices of economic growth in western countries highlight the growing importance of the tertiary or non-industrial sector in creating wealth (indeed the economy of the UK continues to prosper while its industrial base shrinks).

However, the danger in engaging in an abstract discussion of welfare and 'costs' is that actual monetary *losses* (as opposed to expenditure – see Preface) may be confused with the aforementioned intangible losses, which cannot be measured.

'Costs', as we shall see, might simply represent expenditure on noise control devices. To add to the difficulties, monetary losses to individuals or to the country as a whole through absenteeism, sickness, and so forth, are more easily talked about than actually quantified. One cannot simply quote a compensatory figure (whether or not compensation is paid out) for a man who suffers from a certain number of decibels of hearing loss without also taking into account his loss of earning capacity, or the loss to industry through his absence, or the cost to the social services should he be receiving welfare payments. In assessing national costs in this way, it is essential that double or triple counting be avoided. And of course there are more employees who risk suffering from hearing loss, or who already suffer but continue to work, than there are workers actually prevented from working.

We should also bear in mind that sociocusis (hearing loss due to social or environmental causes) is far less common than presbycousis (hearing loss due to aging or medical reasons). As mentioned in the previous chapter, the WHO estimated that health-related costs in America in 1969 amounted to four billion dollars, but this figure was largely made up of costs to the health service (but as this is mainly under private control it represents a boost to the gnp) and compensation payments (which can either be a direct cost to the US government *or* to particular industries). We still do not know the true costs in human or other terms.

The concrete changes in the value of services and property are more precisely evaluated through cost-benefit and related methods of analysis. The assertion by Patrick Rivers that the community around Heathrow Airport loses about £66 million a year[1] from the decline in house values and amenities is based on the fact that estate agents can provide price ranges of houses in areas that can be considered as desirable or undesirable. However, residents themselves have quoted their own prices for quiet housing. According to Rivers, owner-occupiers were asked to imagine that the noise nuisance from an airport was so bad that conversations were interrupted:

> Sixty-three per cent said they would move if adequately compensated. Some would need an average of just on £2,600 minimum compensation, others just on £4,400. The remaining 37%, who would rather stay and suffer noise than suffer moving, reckoned they would need at least £950 compensation each. Even in a situation where people were a few miles from an airport, so that although conversations were not interrupted they would have to strain to hear, 45% would move with compensation ranging from nearly £1,800 to over £3,600 and the 55% stay-puts would need over £570 each in compensation . . .[2]

Academic economists tend to adopt a more testable method to illustrate how the above personal evaluation can be measured. Cost-benefit analysis (cba) is a method of examining the 'trade-off' between advantages and losses, or between utilities and disutilities. A good example of the cba approach is to try to quantify the disutility aspect of aircraft noise on house prices diagrammatically. In Figure 2, the line D^0 indicates the price and the quantity of the housing in a normal, quiet, area; and S^0 the price and supply of houses for sale. After the arrival of an airport, the demand curve shifts to D^1 and the supply curve to S^1. The gap, S^0 to S^1 is believed to represent the price which property owners are prepared to pay to be rid of the noise nuisance, and the shaded box between P^0 and P^1 shows the actual monetary loss through the fall in house prices.

In 1968 this kind of cba was used by economists employed by the Roskill Commission which had been briefed to ascertain the most suitable site for a third London airport. According to the Roskill model, new entrants to the noisy housing zone were assumed to be fully compensated by the relative fall in house prices, i.e., the added value they would get would equal in money terms the loss of amenity due to the noisy environment. Those who sold out because of the noise would suffer a loss as shown in area 2, because their subjective expectations exceeded P^0 but lay below P^1. (It is interesting to note that, in 1970, a Board of Trade study used depreciated house prices, as a measure of the disbenefits of noise – costing around £28.8 million – to compare them with the cost of retrofitting aircraft – £34.9 million.)

One major difficulty in this kind of exercise is that the expectations of people moving into this kind of area are purely subjective. They could either overestimate or (more likely) underestimate the degree of nuisance arising from airport zones, and in any event they will be unable to measure their subjective expectations in the strict financial terms Rivers is talking about. We have already seen that people are attracted to airport areas, in the same way that rural dwellers are attracted to more expensive urban homes, because of the improved prospects of employment.

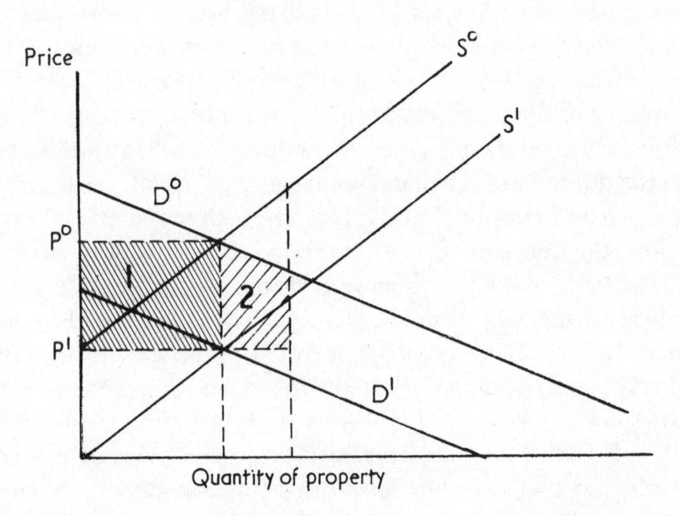

Figure 2 Adapted from Pearce, OECD, 1976

E. J. Mishan points out that, as quiet districts continue to disappear, there is a narrowing of differentials between areas.[3] And, as David Pearce reminds us, if a resident sees a depreciation in the value of his property (in opposition to the usual trend) he would no longer be able to afford a comparable house to move into, and this would inhibit him from selling.[4] Hence, all house-owners in a similar financial situation are preventing the economist from performing his cba in terms of *actual* financial losses, since the cash flow arising from the turnover of houses is smaller than it might be.

These criticisms of cba might appear to be depressingly negative and dismissive. However, we should recognise that all social sciences lack the precision of natural science, because many of the imponderables of social life cannot be rigorously enough dissected and measured. This does not mean that progress in putting a more accurate price on the disamenity aspect of noise will not continue. This is very much a pioneering subject, and its usefulness will grow as we gain a better technical understanding of the effects noise has on people.

Moving away from the more theoretical aspects of assessing the decline in noise-impacted property values, we can turn instead to more concrete calculations of the actual costs of deflecting, controlling or abating noise emissions, and to the payment of special grants, etc., to the sufferers of noise.

Some of the heaviest costs of controlling noise in society fall directly upon the government as it becomes more involved in planning and legislating against community noise, and as it aims to act upon the Industrial Injuries Advisory Council's (IIAC) recommendation that occupational hearing loss should entitle the sufferer to a disability award in line with that of other victims of industrial injuries. But firstly the government, as will be shown more fully in Chapter 11, tries to implement planning and design measures to minimise the noise nuisance at source. Under the Land Compensation Act of 1973, the planning authority in charge of a road building project can buy enough land to help it shield surrounding areas from the noise arising from the construction and use of the new road. This might also mean that it has to erect noise barriers, lay fresh grass over screening mounds of earth, or provide other landscaping such as trees. A protective barrier 300m (980ft) long and 2.6m (8.5ft) wide was erected along the expressway linking London with Heathrow at a cost of £30,000.[5]

However, residents who are exposed to exhorbitant amounts of noise from new construction works, and who feel that it is so bad that they have to take their families away, can get help with all reasonable expenses while they move to temporary accommodation. A householder may also be eligible for compensation for any depreciation in the value of his home if he wishes to sell; the local authority may be willing to buy an owner-occupier's house if the outside noise is so obtrusive that the owner wants to move out. In addition, of course, the local authority will allow insulation grants to certain householders exposed to excessive construction or traffic noise.

We can get some idea of the cost of insulation grants from studies carried out by the Road Research Laboratory. They have estimated that the protection of all British homes exposed to a level of traffic noise in excess of 65dBa L^{10} could cost as much as £1320 million.[6] (The DOE begins to pay for insulation against traffic noise at 68dBa (L10 18 hour) and above.) If ventilation systems are needed (and this is usual when double glazing is installed), it has been calculated that a bedroom or living room would cost between £80 and £130 to gain a noise reduction of about 28dB, and £180 for a reduction of about 38dB. However, the government allows insulation grants of only 60% of the full cost, or up to a maximum of £150 for three rooms. Further, the grant applies only to houses built before 1965. Only about 5% of people eligible have applied for grants because their availability is not widely known.[7]

Part Three of this book is concerned with the practical remedies for the control of noise from all its major sources, and the reader will no doubt come to the conclusion that noise-control is a costly business. Industry will have to invest an enormous amount in retrofitting and in redesigning machinery. The cost of any system used will be high. But we still do not know *exactly* how high. The true costs of noise remain hidden from the expert and the noise-sufferer alike.

For example, independent estimates for the market for noise control in the USA suggest that it will rise from $600 million in 1970 to $1.2 billion in 1980.[8] Other estimates suggest that industry loses $4 billion annually through noise-related problems.[9] According to Bugliarello *et al*, standard diesel engines of 300hp can be quietened by 5 or 6dB at a cost of 1% of the cost of the vehicle. In the USA, in order to achieve the reductions in noise to meet the 1977 83dB limit for all trucks, the cost to the truck purchasing industry would be $34 million, and this will rise to $294 million when a future 75dBa limit comes into operation. A. A. Walters suggests that the cost of retrofitting the entire US-airlines fleet would be about $2 billion.[10]

By and large, the cost of controlling noise results in a net economic loss to those countries carrying out noise-abatement programmes, even taking into account the beneficial impact it may have on various gnp rates. And yet this loss is often judged by influential academics concerned with the disamenities of growth (like E. J. Mishan) to be worth the price paid in exchange for an improved environment. For this reason the subject of noise and costs is set within a complicated frame of reference that discusses not only the increased price the consumer (the man in the street or the public authorities) has to pay for more silent appliances, but also the *human* costs of a social disutility which often have to be weighed against the benefit to society of a public utility, such as an airport.

Although the cost of noise control has not been neglected by acousticians, little real progress has been made in providing universal correlations between certain types of noise sources and the costs of reducing fixed quantities of decibels. In view of the very wide range of noisy machines involved this is not surprising. Many types of retrofit device have been designed solely for use with highly specialised pieces of machinery. As a result, most of the figures that are available are based empirically upon what particular manufacturers have had to pay to retrofit certain types of machines, and extrapolations are sometimes made from such figures in order to present a nationwide picture.

Has government pressure so far offered the only incentive to industry to reduce the noise made by manufactured goods such as cars? Does public and private demand for quieter appliances bring about the necessary supply? Unfortunately, attempts by industry to retrofit or redesign are thwarted as much by the technical difficulties as by the prohibitive costs entailed. If the demand for quieter cars grew significantly, the introduction of a new type of power, such as that provided by gas turbines or sodium sulphur batteries, would be the main aim of motor designers. Indeed, designers have only recently had their attention distracted from the more traditional marketing aims of their employers toward noise-abatement problems.

In the meantime, marketing personnel, when obliged by consumers to state how quiet a particular model is, tend to gloss over crucial technical features that have a bearing on the subject, and which are not fully understood. It is not uncommon for a manufacturer's ratings to show a discrepancy against the actual emission levels when the equipment is finally used.

A study by the US Environmental Protection Agency (EPA) showed that manufacturers of compressors tended to deny any culpability for noise; it was attributed to 'piping systems' beyond their control. More importantly, it was noted that customers were prepared to relax their noise limit requirements in the face of possible steeper costs.[11] A firm might go so far as to provide the service of a noise consultant at no extra cost. But if the customer's needs went further the supplier felt justified in asking that research and development costs be met by the customer himself.

What becomes apparent from all this is that, compared with more conventional consumer demands for efficiency, performance and economy, the demand for silence is significantly low. No doubt the problem of noise will be given more serious consideration by manufacturers under the pressure of official action. In this regard we should bear in mind that the recent availability of lead-free petrol in the USA and the change in carburation and exhaust design of US cars (and British cars made for export) came about because of US anti-pollution measures.

Attempts have been made in the past to quantify the costs to industry of reducing noise to specified levels, largely for the benefit of employees. Or efforts are made to do a form of cost/benefit analysis to ascertain the economic advantages (of improved industrial efficiency) in contrast to the initial costs of making engineering modifications. Inevitably, such attempts are conspicuous in their failure to provide any very reliable data. The EPA admit that studies of the cost of reducing noise – as opposed to air and water – pollution are in their infancy.[12] It is reasonably clear why this should be so.

Firstly, although independent costing studies have been done, they provide merely a very broad guide to the wider economic implications of abatement, in whatever sense this is viewed. Because of the different criteria used in various assessments wide discrepancies emerge, so that quoted figures tend to have an auction-like quality about them. For instance, Charles Wakstein cites a monitoring exercise which revealed that, in the USA, average costs of abatement amount to approximately $70 per decibel per worker.[13] But this contrasts with the Council for Environmental Quality's (CEQ) $26 per decibel per worker.[14] Another estimate suggested that the total yearly cost of quietening machines in industry amounted to only ½% of the US gnp.[15] Almost anything is feasible if enough money is spent on it. Industrial managers are under some official duress to minimise noise in the place of work, so they cannot adopt the same dismissive attitude to the problem as they tend to do in the case of noisy consumer appliances. On the other hand most reasonable authorities are not unaware of the impact of crucial cost-factors on firms operating in a stringent economic climate. What is clear is that there are certain unavoidable expenses involved that industrial managers have to take into account: research performed by institutes, acoustical and absorbent materials, ear-protection devices, masking piped music, etc. Here are a few examples of the relationship between costs and reduced noise-levels.

1 In Chicago an average of $2,600 was spent by each of fifteen factories to decrease site boundary noise levels by up to 23dBa.[16]

2 At a General Electric jet engine plant in Ohio, white-collar workers were protected from an ambient sound level emanating from an adjacent factory of 75–78dBa by considerable structural alterations to the offices (by resuspending ceilings, soundproofing doors and acoustically treating walls). This was achieved at a cost of about $10,000.[17]

3 A paper mill installed a new wood-chipper to speed production processes, which subjected the monitoring operator to levels of 114dB. So a soundproof booth was constructed for the operator at a cost of $2500.[18]

4 A manufacturer spent $10,000 to quieten a large press by something like 25dBa.[19]

5 The manager of Dunlop and Rankin's Steel Service Centre, at Leeds, commissioned a new steel decoiling plant at a cost of £750,000. Noise emanating from the cutting and stacking of steel plate at times reached 119dB inside the factory, and residents complained of an ambient noise level of 56dB. Among the various alternatives considered was the entire rebuilding of the premises in brick and tile at a cost of £275,000; and the removal of the stacking plant to another part of the site, without rail access, at a cost of £60,000. Instead, Dunlop and Rankin called in ICI Acoustics Ltd who manufactured and installed a sound-reduction enclosure for a 61m – 200ft – long machine (the main cause of the high emissions) that resulted in a noise diminution of 77dB inside the revamped plant, and 43dB outside.

Previously the shift-working rota had been reduced to daytime only; afterwards three shifts could be operated. The overall cost was £30,000.[20]

6 Investigations by the University of Salford in the Liverpool area to ascertain cost-effective solutions resulted in the University's own design, manufacture and installation of the necessary acoustic hardware. Much of it constituted silencers of various types:[21]

Cost of silencer	Noise reduction in dBa
£4500	30
£3000	35
£2000	30
£2000	16
£ 200	10
£ 125	12

Once again, we should remember how individualised most of these figures are. Indeed, the absence of industrial homogeneity is probably what inhibits legislators from imposing blanket noise restrictions on industry and economists from arriving at handy universal figures.

The total cost to industry can only be a 'best estimate', because no exhaustive study has been done of each plant affected by noise regulations or even of the majority of industries so affected. The exercise performed by Bolt, Beranek and Newman was of only 68 plants which represented ½% of all workers in nineteen light industries in the USA.

The survey they did in 1973 ascertained that adherence to the 90dBa limit would cost industry $13·5 billion (by extrapolation), and adherence to the more strict 85dBa limit would cost $31·6 billion, a difference of $18·1 billion. However, the EPA criticised this first report for including cost data based wholly or in part on extrapolations from individual plants to whole industries, and from smaller industries to very large ones.[22] To complicate matters, BBN, in a later report in 1975, presented a revised costing schemata which brought the two respective figures down to $10·5 billion and $18·5 billion – the major reason for this was a 'more appropriate' extrapolation to a wider industrial base.[23]

(The bulk of BBN's work entailed the costing of 'monitoring' (record keeping, audiometric testing, etc.) exercises, since these were demanded in recent US legislation. A basic concept used was the definition of impairment used by the AAOO (see Part Two). If this changed, then their figures would also change. Monitoring as required by the government was estimated to cost about $155 million a year ($12 per worker), while audiometric testing for all workers currently exposed to 85dBa will cost about $86 million per year ($20 per worker). Thus the proposed OSHA regulations, according to BBN, will cost a total of $241 million yearly for the purely monitoring aspects.)

In spite of the shortcomings of these exercises, there is still a need for the kind of study performed by BBN. Some of the problems arise from entirely conceptual and definitional phenomena. But others are of a far more mundane character, concerning the number of noise-deafened workers, which forms a basis for studies of the costs of engineering modifications and protective equipment. However, on the available figures (given in Chapter 3) some simple calculations can at least be made about the cost of minimising the impact of noise on the long-suffering worker.

A British manufacturer supplies earplugs and muffs in a range of three qualities. Discounts are available for quantities up to ninety nine, and further negotiable discounts for greater amounts. The cheaper discounted version is sold at £2·49, and is the most widely used in industrial concerns. Two more expensive versions cost £5·16 and £5·53, and they are generally used by airport workers, power generating station employees and by others in 'high noise industrial situations'. Foamed polymer earplugs are also sold in batches of 200 at 18p a pair, and possess good noise attenuation characteristics. Indeed, it is theoretically unnecessary to use muffs until levels of over 100dB are reached, as the much cheaper plugs will afford adequate protection in all but the heaviest industries, as Figure 8 in Chapter 9 indicates.

So using the Factory Inspectorate's figures of 1,161,000 workers exposed to 90dB or more daily, and assuming that the cheaper version of the muff is provided to all workers nationally, we arrive at the figure of £2,890,890 as an overall cost to British industry. If we take the lower decibel figure of 85, the cost rises accordingly to £5,149,320. Taking the US Department of Labour's 1974 figures of 8½ million workers exposed to 85dBa or more, the cost to US industry is $332,860,000 (£173,365,000).

Alternatively, if we assume that earplugs, instead of muffs, can quite as effectively minimise the impact of 85dBa or less, the cost to industry would be much lower. For example, to protect 2,068,000 British workers exposed full or part time to 85dBa or less means that the supplying of polymer earplugs would cost only £228,240. However, if the costs are viewed longitudinally, the need must be considered to replace muffs and plugs annually.

In conclusion, it is clear that, from one perspective, the human and other intangible social costs have to be separated from the costs of abatement. And the expense of quietening consumer goods is different from the cost of minimising occupational noise, both in motive and in methods.

The real significance of the questions of noise-costs centres upon issues of ethics and economics, and the latter discipline can help us better to work out a rigorous system of the former. The question can be reformulated as 'Does noise pollution have an economic impact upon the nation, and is the price worth paying?' Manufacturers of consumer appliances need not make a loss, since the cost of retrofit devices can be added on to the retail price; the cost of controlling noise in factories will be partly offset by less absenteeism and improved efficiency. The remaining substantial losses can be written off as the price the nation has to pay to abate an insidious and harmful hazard.

However, the future usefulness of cost-benefit analysis is limited. It is questionable whether cba should be used to resolve some of society's most crucial decisions; to decide, for example, how to achieve a balance between individual rights and public disamenities, as apparently the Roskill Commission tried to do. The planner himself is often apprehensive about making such decisions. But the economist, even with the aid of cba, is also uneasy about the fact that his conclusions may be used as a substitute for political decision-making. At the moment, as David Pearce suggests,[24] we must be content with merely refining and sharpening our tools so that at least future generations might have a better idea of the *true* cost of noise pollution to society.

Part Two:
THE TRANSMISSION PATH OF NOISE

5

NOISE AS AN ENERGY FORM

The energy properties of noise are remarkably versatile, penetrating and insidious. Noise can pass through water, solid objects, building materials such as concrete, and most gases in addition to air, which is the most highly elastic medium for noise transmission (it cannot pass through a vacuum). The energy of sound is transmitted *via* millions of colliding molecules of air. The actual speed of transmission may be represented as a series of compressions and rarefactions within these air molecules, which travel away from the source of the sound.

The range in power of the sounds encountered in everyday life is extremely wide. For example, the noise of a jet aircraft is 10^{13} times as intense as that of a whisper (about 20dB), or thirteen increases of 20dBs in tenfold units. A tenfold increase is called a Bel (after Graham Alexander Bell), and it is the unit of measurement used to describe a particular kind of acoustic energy ratio.

The range of sound *intensities* even exceeds that of sound pressure as it increases at twice the rate. The doubling of a sound pressure would take four times as much energy to increase the rate of oscillation of air molecules to achieve the same doubling of intensity. Surprisingly, a doubling of sound intensity yields an increase of only 3dB, because decibel levels are progressed *logarithmically*, rather than arithmetically, and the logarithm of a doubled number to two places of decimals will always go up by 0·3 (a doubled intensity level (I.L.) will hence go up by 0·3).

If we take two sound powers S^1 and S^2, where $S^2 > S^1$, then the number of Bels (B) is given by the formula:

$$B = \log \frac{S^2}{S^1}$$

(if S^2 was less than S^1, B would have a minus sign).
So using the example of jet aircraft and the soft whisper:

$$S^2 = 10\text{kW}, \quad S^1 = 0{\cdot}001S.$$

Converting to watts:

$$S^2 = 10^4\text{W and } S^1 = 10^{-9}\text{W}.$$

Therefore

$$\log \frac{S^2}{S^1} = \log \frac{10^4}{10^{-9}} = 13.$$

For practical purposes the Bel is too broad a unit; it is therefore more convenient to subdivide into decibels, where 10dB = 1 Bel. So our formula now becomes:

$$\text{dB} = 10 \log \frac{S^2}{S^1}$$

i.e., doubling the sound power level gives a 3dB increase.

The implication of the logarithmic progress of decibel levels is that only a small range of numbers is used to measure a wide range of sounds. Hence the total range of sound levels from the threshold of hearing to the threshold of pain (0 to 140dB) represents a sound-pressure range of 1 to 10,000,000. In other words, the threshold of pain is 10^{14} times more intense than the threshold of hearing. As an illustration, ordinary speech is measured at 60dB, street traffic at 70. However, street traffic is not 10% louder than speech, but rather *10 times louder*. So a relatively minute change in decibel value can mean a tremendous change in the intensity of noise. In both the social and industrial environment, any improvement that causes even a small decline in the decibel level is significant. A drop of 3dB means that the noise level has been cut to half its previous level.

The compressions that sound-energy enforces on dense air-molecules create pressure fluctuations. The *rate* of these fluctuations is known as the frequency (or the number of completed vibratory cycles per second, c/s), and the internationally preferred unit of frequency measurement is the Hertz (Hz). *Amplitude* refers to the magnitude of the pressure fluctuations.[1] In comparison with atmospheric pressure (around 1000 millibars), miniscule fluctuations in pressure are needed to produce an uncomfortably loud noise (about a tenth of a millibar).

It is not generally appreciated just how inefficient is the generation process of energy in terms of its conversion into sound. In the home a 150-watt refrigerator generates a nearly inaudible 40dB at a distance of about 1m (3ft) which is equivalent to a tenth of a microwatt of sound.[2]

Table VI: A sound at the threshold of hearing has an intensity of 10^{-12}W/m^2. The sound intensity factor shows how many times more intense a given sound is than a sound at the threshold of hearing. Courtesy of Atlas Copco Ltd

	Sound intensity factor		Sound intensity level (dB)	Source
	100 000 000 000 000	injurious range	140	jet engine (25m)
Threshold of pain	10 000 000 000 000		130	rivet gun
	1 000 000 000 000		120	propeller aircraft (50m)
	100 000 000 000	danger range	110	rock drill
	10 000 000 000		100	metalworking shop
	1 000 000 000		90	heavy lorry
	100 000 000	safe range	80	busy street
	10 000 000		70	private car
	1 000 000		60	ordinary conversation (1m)
	100 000		50	low conversation (1m)
	10 000		40	soft music
	1 000		30	whisper (1m)
	100		20	quiet town dwelling
	10		10	rustling leaf
Threshold of hearing	1		0	

Noise, therefore, clearly possesses some curious energy ratios. Loud noises can be produced by small pressure fluctuations; yet a person shouting is unlikely to have his vocal form of sound-power actually exceeding a thousandth of a watt. Similarly, energy dissipated as noise is a minute part of the total energy output of any industrial unit, since it is difficult to conceive of noise as a wasteful byproduct of production processes. On the contrary, its suppression, even in quite moderate amounts, results in considerable losses of energy. For many types of machine, something in the order of an energy loss of a thousandth of 1% applies. The economic considerations are not in the same league as those concerning the loss of fuel through inefficient combustion processes.[3] Again, should the noise emissions of a jet engine be converted into measurable energy waves they would only produce a few tens of kilowatts, compared with the total power of the energy itself which may produce as much as a million kilowatts.[4]

Hence the most practical methods of noise control attempt to absorb or interrupt the flow of acoustic energy. Such methods are very expensive, because the problem is not merely one of containing the equivalent of tens of kilowatts of energy. Human hearing can only deal with very small amounts of sound energy and second-order physical processes which cannot easily be controlled or modified. Moreover, to reduce the noise of a jet engine to a tolerable level, a reduction in the conversion of *kinetic* energy into sound from one part in ten thousand to somewhere around one part in a million would be needed.[5]

Strangely, because of the logarithmic progression of noise measurements, a 10% decline in noise is almost completely unnoticed by most people. To the industrialist, however, such a decline would be conspicuous in its energy loss and consequent rise in costs. In terms of objective measurements in the physical world, quite large steps would need to be taken to correlate with the subjective measurement of perceived noise. For instance a 10% decline in noise levels is detectable by the ear only with difficulty; a 50% decline is noticeable but unremarkable, and a 90% reduction approximately halves the subjective impression of noise. It is for this reason that the reduction of subjective noise levels is such an unrewarding activity; any reduction must exceed 10% to be significant.

For example, in an instance where noise from an industrial installation disturbed residents in a rural area, the acoustic power crossing the boundary was reduced by a factor of thirty at a cost exceeeding £100,000. Although this was accomplished by using standard noise reduction techniques, this particular industrial unit succeeded only in reducing the ambient level to about 50dBa, whereas the residential community were anticipating a noise level of 40dBa.[6] To have diminished the level by a further 10dBa would have meant lowering the noise output to 1/300 of the original output. Such features in regard to the nature of aural energy perception must be taken into account in regard to the practical implementation of noise reduction techniques in industry.

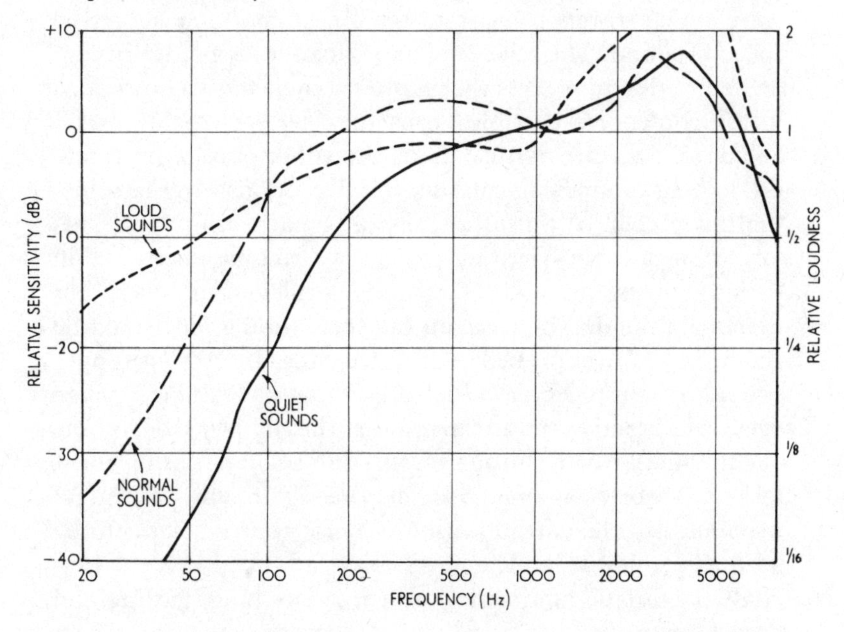

Figure 3 Hearing Sensitivity Curves. Courtesy of T. A. Henry, University of Manchester, 1976

The fact that noise is a form of energy is amply illustrated by the internal combustion engine which, as Rupert Taylor has suggested, is a device for converting noise into mechanical power, since explosive gases are used to thrust pistons down a chamber.[7] All machines are to a greater or lesser extent inefficient: the IC engine in fact releases some of its combustion and other mechanical noises into the environment.

However, all non-explosive noise emanates from the creation of air-pressure perturbations arising from the movement of solid surfaces. Apart from this initial disturbance, noise also arises from the modification or amplification of the disturbance and the radiation of sound. Whenever air moves past a solid object or vice-versa, there is a vortex formation: the sudden reduction in air-pressure in a moving object's wake causes turbulence, and thus noise. There is also pressure on both sides of a moving object, and this constitutes a sound wave which travels away from the object continuously at the speed of sound. The science of aerodynamics is naturally concerned to design objects that move through the air so that as little turbulence as possible is caused.

The air pressure perturbations can either pulsate (to create fluctuating volumes) or can slice through the air (to create fluctuating forces).[8] A pulsating pressure emanates from a transformer, or a large vibrating panel, or a ringing anvil. A thin surface cutting through the air, and displacing no mass, but exerting a fluctuating force on the surrounding air, would be typified by a fan or propeller, or any moving object whose opposite sides move together in such a way that no 'volume' change occurs. In practice, industrial noise can be composed of any one or combination of these sources. It is thus seldom possible to establish theoretical methods of prediction on individual machines, so the categorisation of noise sources ought to be used as a guide only. The most noisy industrial processes include pneumatic hammering, riveting, chipping, fettling and drop-forging.

Apart from the obvious noise source of the cylinder block and crankcase (which is considerably contained by the cast-iron housings), motor-vehicle noise arises from resonant timing covers and other moving components, exhaust outlets and tyres. Vibration from the tyres is transmitted to the vehicle interior through the body mounts, steering column and suspension system. A fast-revving small engine will always sound noisier than a slow-revving, large engine, which will also emit noise of a lower and thus less irritating frequency. For this reason, large US cars are responsible for more air, rather than noise, pollution; in the USA, the noise irritant mainly arises from the contact between wheels and road surface.

Diesel-engines are noisier than their petrol-powered counterparts because of the more noticeable airborne vibration, especially when they are loaded. The petrol-powered engine also has a more even spread of combustive power throughout the chamber, whereas the diesel relies on a much higher compression of air-to-fuel ratio in the fuel-burning process, so that the explosive power is more instantaneous.

All jet engines develop their thrust from a stream of high-velocity hot exhaust gases being ejected from the tail-pipe. These gases are turbulently mixed with the surrounding cold, atmospheric air which is drawn into the engine to create a vortice, which is then passed over a turbine used to drive the compressor. It is the compressor that creates the screeching high-frequency sound, and complements the low-frequency roar.

The basic jet engine is the turbojet, which has been modified and is now known as the turbofan jet. Turbofans are used on the present generation of subsonic aircraft such as the Boeing 707, the VC-10 and the Trident.[9] The principal difference with the turbofan is that after passing through a low-pressure compressor, the airflow is divided; about half passes through a high pressure compressor; the remainder is bypassed and is either exhausted as a separate cold jet or is blended with the hot jet before passing through the final nozzle.

The turbofan has the advantage of improved propulsive efficiency in a subsonic cruise flight. On the more recent wide-bodied subsonic aircraft, such as the Tristar, the Boeing 747 and the Lockheed L.1011, the bypass principle in the engines has been taken further. Only about a sixth is compressed, burned and exhausted as the hot jet stream. The remainder is exhausted directly after passing through a large fan system (i.e., they have a high bypass ratio), and provides the major proportion of the total engine thrust.

The high-bypass engine is hence less noisy than the low-bypass. The Tristar achieves a noise certification level of 15 perceived dB lower than the older type of aircraft. However, high-bypass ratios are not used in supersonic machines because they are basically less efficient at supersonic speeds. The Concorde with its Olympus-593 engine must have a high wing-loading and sharp sweep-back, which demands a much greater installed thrust than a subsonic craft, to take off correctly. Consequently the thrust-to-weight ratio must be 0·4, compared with 0·22 and 0·29 for subsonic aircraft.

In addition to jet or fan noise, industrial machines are characterised by the use of mechanical energy to reshape some piece

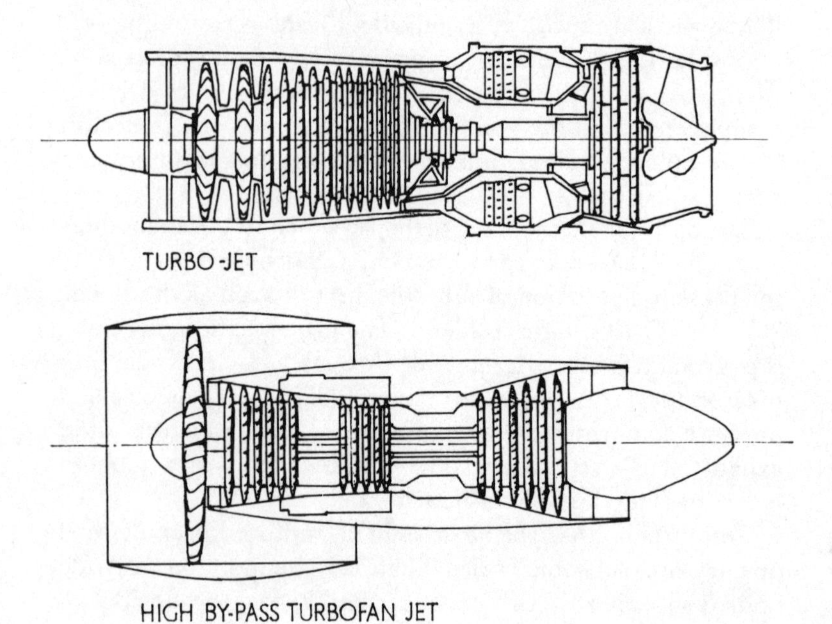

Figure 4 Source: *Noise*, Rupert Taylor, Pelican Books, 1975

of material (e.g., pressing) or to take parts out of it (e.g., machining), or to reconstitute it in some other form (e.g., die-casting). The energy levels needed in most instances are impulsive, and are noticeably common in work involving sawing, planing, punching, stamping, pressing and riveting processes, all of which have varying levels of impulse and repetition rates.

E. J. Richards lists six categories of processes producing noise:

1 *hammer deceleration*, where an energy device accelerates or decelerates on impact, giving rise to pressure waves;

2 *workpiece distortion*, where changes in shape cause pressure waves;

3 *anvil ringing*;

4 *supporting structure noise*, where resonance causes pressure perturbations to continue to oscillate for several more seconds;

5 *air ejection*, either aerodynamic or acoustic; and

6 *blow-off valve noise*, frequently where mechanical parts are put into reverse.

Vibrations are also caused by the rotation of unbalanced components, like the weight on a rotating arm. When a machine is oscillating in one mode, say up and down, the displacement that causes the vibration will be a 'pure tone', whereas other machines that vibrate in other directions create more complex, and more prominent, vibration forces. A dilemma for the designer is that the accelerating forces which create vibrations when they pass into foundations may not be amenable to balancing out measures of control.[10] In practice, there are two types of vibration that cause the most problems; *forced* and *self-excited* vibrations. The former occurs when a simple harmonic disturbance directly affects a machine element. The latter occurs when elements themselves give rise to vibrations.

Specific types of industrial noise may arise from installations in the following manner:

Furnaces: from the airborne noise radiated from burners and the furnace walls, and from flue gas ducts and stacks;

Burners: from turbulence generated in the air being drawn in through the air register and by the combustion process;

Boilers: from forced draught fan and ducting, and from low frequency rumbles;

Electric motors: from the cooling fan, or from powerful tonal sounds in magnetic components;

Steam turbines: from tonal noise arising from blade-rotation; and

Gas turbines: from the harsh internal environment caused by the exhaust and gas stream.

The range of sound pressure levels from a wide variety of industrial plant can be of quite a high order; Table VII gives some indication of the levels that can be reached.

The actual impact of industrial noise sources on the employee himself does not necessarily correspond arithmetically to recorded decibel levels. A US power-generating plant, using equipment with very high noise levels, was said to produce a 'minimal' impact on the work environment because equipment such as furnaces, switching-stations and turbo-generators were remote-controlled.[11] Similarly, noise sources at a typical US car-assembly plant varied in impact from 'minimal' to 'considerable'. Many of the noise emissions at some work areas had been reduced to below 90dBa, but no such structural containment of noise could be achieved at areas like the rough-grind booth, where ear-defenders had to be worn.

Table VII: Noise levels in industry.
Source: E. J. Richards, University of Southampton, 1976

Classification of machine or process	Noise level dBA
Air cleaning bench for machine castings	109
Blower for heating of blanks	104
Chain riveting machine	111
Chipping hammer	118
Chisel, pneumatic	116
Dicing machine	100
Diesel generator set	110
Dust collector plane	98
Feed slide for tins	100
Forge plants	107
Hammer (pneumatic)	122
Handling & transport of billets	98
Mill-alumina	109
Mill-coal	107
Air-line for cleaning down	102
Ball mill for powderizing materials	106
Concrete pipe making machine	97
Crusher plane in quarry	110
Die casting machine	97
Drop hammer (peak)	111
Fan (for tank cooling)	97
Fettling tool	104
Grit blasting machine	104
Hammer (for peak plate levelling)	110
Lathe (wheel)	108
Linishing machine	104
Mill (flint)	101
Mixing Plant (plastics) (peak)	110
Mould machine in foundry	101
Nut runner (air operated)	106
Perforating machine	101
Power hammer	108
Press—cropping	100
Press—laminations	98
Wet riddle	106
Woodworking circular saw	104
Woodworking double end tenoner	100
Nailing machines	95
Peening gun	97
Plasma arc profiling machine	110
Saw bench—plasterboard	106
Saw—friction	118
Saw—for batches of tubes	105
Saw—slotting	112
Woodworking planer	103

6

NOISE, HEARING AND WELL-BEING

The ear is a highly sophisticated and sensitive organ that can analyse sounds across a range so wide that the highest frequency is almost a thousand times as great as the lowest. And it is ingeniously engineered so that the advantages of such fine tuning are made to greatly outweigh the disadvantages: the membrane of the ear has an averaging effect to prevent such small pressure signals (of about one ten thousandth of a millionth of atmospheric pressure) from being swamped by the random thermal motions of air molecules against the ear drum. The two separated ears enable the brain to identify the direction of a sound, by assessing the differences in the time of arrival of the sound at each ear.

One of the first observations made by acousticians is that the loudness of a sound as sensed by the brain is not proportional to its measured intensity, as it doubles for each tenfold increase in intensity. In order to devise and calibrate meters that can measure sound commensurately with the ear, acoustics experts need to be aware that the ear perceives fluctuations in the range of 20–16,000Hz as continuous sound. Human voices contain most energy in the range of 500–3000Hz, so it is not surprising that the ear is most sensitive in this range. In effect the ear discriminates against low frequencies, making these appear relatively quiet. The sensitivity drops away above and below these levels, and allowance is made for this non-uniform sensitivity when noise levels are measured. Most sound-level meters therefore have built-in filters that imitate the ear by attenuating the lower frequencies, thereby obtaining physiologically realistic noise-level readings in this way.

The chain of events which leads to the sensation of hearing starts with the passage of sound waves down the ear canal, which extends from the outer ear to the ear drum (tympanic membrane). The outer ear (or the auricle) performs the often unrealised function of softening the abruptness of the change from free air to air that is funnelled down the ear canal prior to its arrival at the ear's sensory mechanism. Dust, insects and other unwanted bodies are kept out of the canal by hairs, and by the wax that is formed to trap these natural enemies.

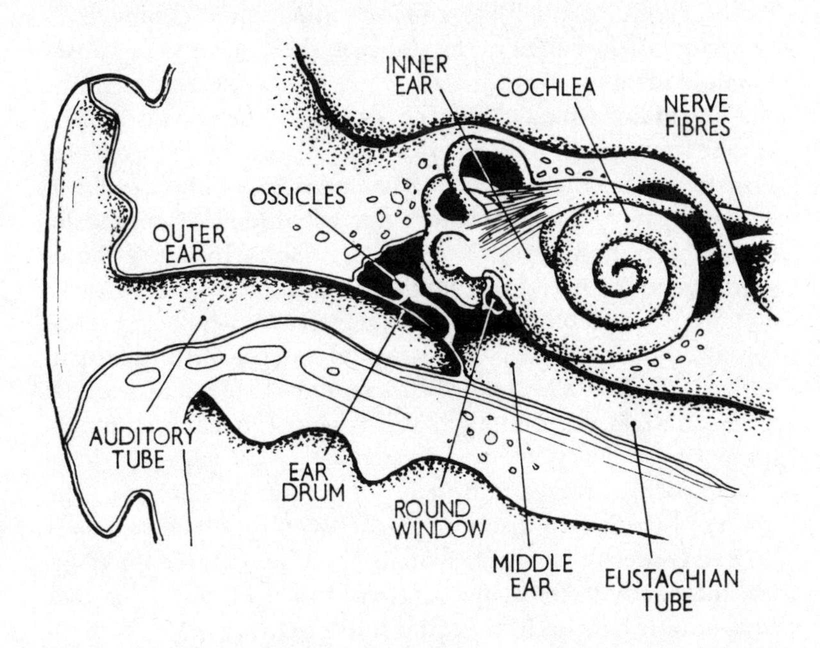

Figure 5 Cross section of the Ear. Source: V. A. Hines, *Noise Control in Industry*, Business Publications Ltd., 1966

The ear drum of course vibrates when the air's pressure fluctuations impinge upon it. It is merely a diaphragm kept under slight tension by muscles known as the tensor tympani and stapedius. Attached to the ear drum is the middle ear, which is an air-filled cavity containing a chain of three small movable bones. The latter are known as the auditory ossicles.

The middle ear is also an impedance device, and is in a sense an acoustic transformer. The movement of the ear drum is transmitted mechanically by the ossicles to the inner ear, which is a complex system of fluid-filled cavities positioned deep in the skull for maximum protection. The system of cavities includes the cochlea and three semi-circular canals at right angles to each other (these canals also contribute towards controlling the body's sense of balance).

The cochlea is considered to be the most complex part of the ear. Its minute snail-like coils contain a liquid known as perilymph. Movement of the stapes in the oval window causes the fluid to vibrate with the assistance of the membrane of the round window which allows the pressure to be equalised. The actual transmission of this vibration to the brain is accomplished by a partition of two further membranes in the cochlea, between which are 24,000 hairs connected to nerve endings in the auditory nerve system. It is these hairs which respond to vibration in the perilymph fluid. And it is the connected nerve endings which send to the brain the messages which create the sensation of sound as a phenomenon external to the human body itself. This is perhaps the most remarkable quality of the ear; sound is not simply heard within the ear, its multi-dimensional reality is also understood.

The cochlea is in a sense a crude frequency analyser. The ability of the ear to discriminate between different sounds is believed to result partly from analysis of the sound in the cochlea and partly from analysis in the brain. One theory is that each minute hair (or fibre) of the auditory nerve has its own individual tone to which it is able to respond when appropriately excited. The four rows of hair cells may also be of differing amplitude–sensitivity, and the most sensitive are damaged primarily by excessive noise exposure.

Recent research has shown that a considerable proportion of the outer rows of hair cells can be destroyed before any reduction in hearing acuteness becomes apparent. The ear also has a remarkable built-in 'defence mechanism': if it is subjected to a noise of more than about 90dB lasting longer than about ten milliseconds, a reflex action tightens up the tensor tympani and other mechanical parts of the middle ear. This results in a reduction in sensitivity to low and middle frequencies. Some people are able to spontaneously activate this 'tightening up' process when forewarned of an impending loud noise.

The average value of the threshold of hearing for a large group of otologically normal subjects between the ages of eighteen and twenty-four has already been determined by experts. This value is incorporated in British and international standards, which are roughly equivalent. In view of the ear's subjective reaction to different frequency levels, 'loudness level' is measured in *Phons*, and the sound is compared again to a standard reference signal of 1000Hz. The loudness level in Phons of any sound is taken as that which is subjectively as loud as a 1000Hz tone of known level. For example, 0phon is 0dB at 1000Hz, and 40phon is the loudness of any tone which is as loud as a 1000Hz-tone of 40dB.[1]

The frequency/sensitivity syndrome of the ear necessitates a knowledge of the way in which sound is distributed throughout the frequency spectrum. This can be gained by dividing the noise into octave bands and measuring the sound pressure level (spl) of each band. (An octave is a doubling of frequency so that the range of 90–180Hz is one octave, as is the range 1400–2800Hz.) The octave bands are usually identified by their geometric centre frequencies. For example, the geometric centre frequency of the octave 90–180Hz is approximately 125Hz. Octave-band analysis is used to assess hearing hazards and to specify ear-defenders; it may also provide enough information for engineers concerned with noise reduction. Occasionally more detailed information is needed about the frequency spectrum, when the spectrum itself can be analysed into narrower bands.

Noise scales have, over the years, been devised to measure various subjective degrees of loudness or annoyance. Many different scales have been evolved since the first noise scale was used in New York in 1929 for an experiment on city noise. They all have the advantage of a predictive power in the matter of group responses to the particular type of noise that is studied while devising the scale.

The A-weighted scale forms the basis of most noise-assessment procedures. This scale is believed to ensure the rating of noises so that they relate more to the effects of noise on hearing. It represents an overall measurement of the whole range of audible frequencies made with a meter incorporating a selective filter which can be used to make the instrument less sensitive to the high and low range, and more sensitive to the middle range of frequencies.

Several studies in the USA have been conducted to evaluate the efficiency of using A-weighted sound levels in rating hazardous exposures to noise. In a study of 580 industrial noises it was demonstrated that this level indicated the hazard to hearing as accurately as did limits expressed as octave-band sound-pressure levels in 80% of cases, and was slightly more conservative than octave-band measures in 16% of the cases.[2]

The A-weighting was originally intended to be used at higher sound levels, and was designed to correctly read the response of the ear at levels between 55dB and 85dB. However, as a simple weighting network cannot precisely simulate the highly complicated response of the ear, other networks have been devised and designated by the letters B, C and D. The characteristics of the A, B and C weighting networks are shown in Figure 6.

The A-weighting is normally used for industrial noise measurements. The use of the C weighting derives from its almost uniform frequency-response to noises, so that it can often be used as a substitute for the overall spl. The D weighting has been proposed specifically for aircraft noise measurements, but has not yet been incorporated into national standards. The perceived decibel-rating scheme is also used far more widely in assessing the impact of aircraft noise.

A variety of indices have been devised in the past to help quantify and categorise the disturbing effects of noise on the average person. In the USA there is the composite Noise

Rating, Germany has the Stoerindex and France the *Indice de Classification*. They all combine noise, number and duration to

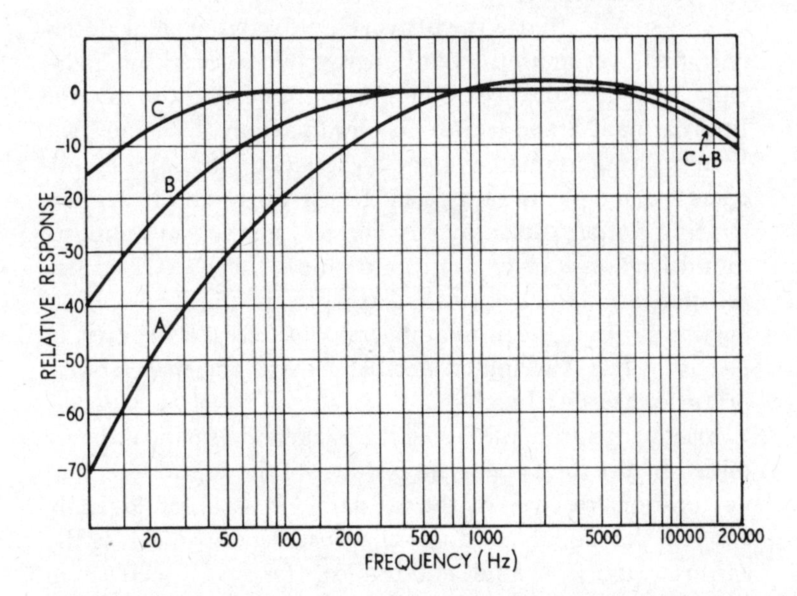

Figure 6 The Internationally standardised weighting curves for sound level meters. Courtesy: Institute of Sound and Vibration Research, University of Southampton, 1975

predict a nuisance index. At present there are four well-known British indices:

1 *CNL* (corrected noise level). This index is specified in BS 4142 and is used to assess the nuisance-value of industrial noise. Corrections are applied to the measured noise-level, to make allowance for some of the subjective facets of annoyance, by adding 5dBa for the impact of an annoying frequency.
2 *Leq*. This is an important index, mainly used for industrial noise. It defines an acoustical quantity dependent on sound-level and duration of exposure. This quantity is referred to as the *equivalent continuous noise level* (Leq of ECNL) over an eight-hour, or typical, working day. 'Noise dose' is often used to

denote a quantity depending on sound-level and duration of exposure, but as yet there is no internationally agreed definition or terminology.

The specification in the British Code[3] is based on the internationally recognised 'equal energy hypotheses' (or 'principle'), as are the assessment schemes advanced by the International Standardisation Organisation (ISO) and the British Occupational Hygiene Society (BOHS). This equal-energy principle is said to be based on research demonstrating a more 'scientifically correct' method of handling impulsive noises (such as drop forging).[4] The NAC has also recommended the Leq system, and prefers a unified scale as an alternative to the present diversity of scales. It believes, in fact, that the A-weighted decibel scale is the most appropriate for use with Leq.[5]

Briefly, the principle is that the hazard to hearing is determined by the total energy (a product of the sound level and its duration) received by the ear daily.[6] It assumes, logically enough, that equal quantities of acoustic energy entering the ear are equally harmful, and it allows for a 3dB increase in sound pressure level for each halving of the duration (below eight hours) of continuous exposure. It has been suggested, however, that the equal energy concept is rather too hypothetical because of the lack of empirical verification in the workforce.[7]

3 *L10.* This is used to assess the impact of road-traffic noise. The most noticeable feature of the general level of traffic noise is its variability with time. In other words, people who live in homes adjacent to public thoroughfares and highways experience degrees of annoyance which vary with the peak noise-levels. The L10 index is generally known as the 'ten per cent level', since it is the level that is measured in dBa and which is exceeded for 10% of the time when measurements are taken (i.e., an L10 level of 70dBa means that the level of 70dBa was exceeded for 10% of the recording period).

The NAC recommended the use of the index L10 (18-hour scale) using the average hourly L10 value between 6am and

12 midnight on weekdays, and this was adopted by the government. Some relevant examples of L10-18-hour values are given in Table VIII.

Table VIII: Some relevant examples of 10–18-hour values.[8]

L10 (18-hour) dB(a)	Situation
80	At 3m (10ft) from the edge of a busy main road in a residential area, average traffic speed 30mph, intervening ground paved.
80	At 18m (60ft) from the edge of a busy motorway carrying many heavy vehicles, average traffic speed 60mph, intervening ground grassed.
70	At 18m (60ft) from the edge of a busy main road through a residential area, average traffic speed 30mph, intervening ground paved.
60	On a residential road parallel to a busy main road and screened by houses from the main road traffic.

4 *NNI.* This uses a similar concept to the L10, and is applied to evaluate aircraft noise. Annoyance depends on the peak noise levels from aircraft flying overhead and on the *number* of them. The NNI measure was hence based on the fact that the subjective reactions of people – as discovered from replies to questionnaires – were found to correlate well with a particular mathematical combination of the average peak ground-level noise and the number of overhead flights. Its usefulness is said to arise from the fact that once the actual traffic flow is known, its nuisance value can be calculated without the need for further cumbersome social surveys, and it enables predictions to be made of future levels of nuisance.

The *limitations* of the index were admitted as early as 1963 in the Wilson report. The original mathematical combinations were derived from a survey of 1,731 people who were habituated to aircraft noise, and it is recognised that there is a wide variation in individual degrees of sensitivity to aircraft noise.[9]

There are other noise indices. PNdB (perceived noise level) is a scale which is designed to correlate better with the subjective perception of aircraft noise, and takes into account intensity and frequency. It is based on the assumption that annoyance approximately doubles with each increment of 10 PNdB.[10] So the annoyance of an aircraft yielding 110 PNdB will appear to be twice as noisy as that generating 100 PNdB, and four times that of an aircraft that shows a recorded level of 90 PNdB. Finally, the Speech Interference Level (SIL) indicates that (excessive) level that seriously impedes speech communication and the Noise Criteria (NC) represents a subjective level that is acceptable to people in buildings.

Agreement is being sought amongst various bodies and authorities for the adoption of specific scales on an internationally acceptable basis. In the meantime, current recommendations may still range from the simple overall recommendations of fixed decibel limits such as 90dB for factories and 68dB for offices, to specific indexes which are often, but not always, time-based or time-averaged. But it is not always possible to establish correlations between different units. Noise Criteria Curves or Noise Rating Curves are still held to be amongst the most satisfactory methods of analysis for offices where speech is normally a vital factor.

Whether a person can properly hear human speech in his everyday environment is a significant criteria in judging the disamenity qualities of noise. The American Academy of Ophthalmology and Otolaryngology (AAOO) considers that a person should be able to perceive 90% of spoken speech in quiet surroundings, although Atherley, Noble and others have found that there is no way of satisfactorily estimating speech and understanding from an audiogram.

Purely technical assessments of sound cannot be adequately correlated with the sufferer's subjective interpretation of the impact of noise on his hearing. In any event hearing tests made without a background noise level (which is common in industrial and social life, and tends not only to mask some louder noises but also to contribute to the overall ambience so that it is more difficult to 'pinpoint' particular noises) are empirically unrealistic. BS4142 in fact called for an established relationship between background noise-level and a calculated noise-level criterion.

In Chapter 1 we saw how harmful noise can be, and we have seen above how systems of noise measurement have been devised. So measuring the harmfulness of noise is the logical next step derived from the noise indexing schemes. All the above systems, even though subjective variations are taken into account, are based on what Burns and Robinson call the *parametric* approach, which assumes that all human hearing processes are fundamentally the same. So this process can be scientifically measured.[1] Once a given noise is described and its duration has been charted, specific degrees of hearing loss can be predicted, at least statistically, along a decibel scale. A formula can even be devised to enable an acoustician to say in theory that a man at work, exposed to 120dB for about two hours a day, would stand a one in four chance of suffering a hearing loss of 48dB at 4000Hz (which is quite severe).

However, as Burns and Robinson remind us, the longest case histories of noise exposure may tend to be associated with persons in better states of health. Some workers may have had better hearing to begin with. Similarly, sociocusis is to some extent a self-limiting condition, as it prevents progressive and continued loss. There is some difficulty in accurately measuring the effects of noise durations for less than an entire working day. Erratic time patterns and the variability of noise characteristics during noisy periods are not easily brought within the experimental requirements of the acoustician. The clinical picture of age- and noise-induced deafness is not identical, and cannot be wholly separated. The range of hearing levels, for example, increases with age. By middle-age, a population of several hundred people would be able to span all the values of an audiometric expert's scale.

An important consideration, relevant to both noise reduction practices and official policy, is whether the *emission* or the *reception* of noise should be the criterion for assessing harmfulness.[12] The expression 'noise exposure', as used in the British Code of

Practice, denotes the noise which a person is subjected to, whether or not protection is worn, although the 'limits' it specifies are for exposure only when the ear-protection is not worn. The Code, of course, also indicates levels above which certain action should be taken.

Plant managers, who think conceptually in terms of noise-emission, do impose certain work-area noise-limits within the broad range of 80–90dBa, and may allow up to 95dBa in areas where personnel are not permanently stationed. These are often designated as 'noise hazard zones'.[13] Therefore, in practice, there is little juxtaposition between the emission/reception concepts. Any reasonable plant manager who has to come to terms with differential exposure-patterns and special hazard areas would impose a blanket limit of, say, 85dBa, or, instead, would insist that ear-protection be used at all times.

Although the physical characteristics of noise are highly variable, each sound can be expressed and recorded precisely as a *wave form* (or *crest*). This represents both its immediate sound-pressure and its length in time. It is difficult to measure because instruments have inertia characteristics which cause a delay in responding to the beginning or end of a stimulus. Therefore, in practice, measurements are made of the average noise intensity over a typical time interval appropriate to the kind of machinery being tested.

A characteristic of noise is the peak, or highest, pressure or intensity reached during a specified interval. These high concentrations of noise-energy are important in terms of their ratios to the average sound-pressure level on the ear. Any noise wave-form can be analysed mathematically into a spectrum of pure tones ranging from zero to infinity, but interest is generally focused upon those frequencies that have an influence on human health.

Certain subjects may show that their maximum hearing losses occur at 3000Hz, and some at 6000Hz, although records show that maximum losses rarely occur at levels beyond 6000 to 8000Hz. High frequencies are more dangerous than low ones at the same pressure level.[14] Intense *low* frequencies can cause maximum hearing losses over the 500–2000Hz range. The subject may not become aware of any affective loss of hearing until the frequency associated with speech is involved, i.e., the 500–3000Hz range. Glorig *et al*[15] studied men and women not occupationally exposed to noise, and compared them with other groups. They found that even without any high-level exposure or ear disease, loss of hearing occurs first and predominantly in the region of 4000Hz.

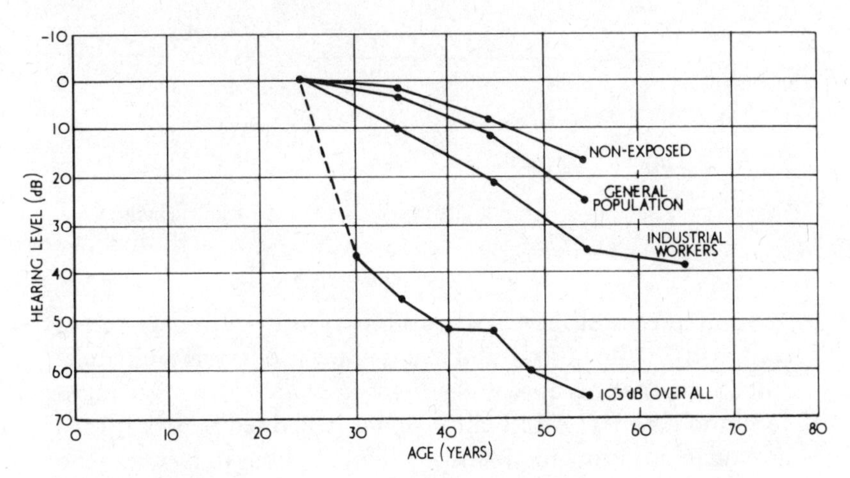

Figure 7 Median Hearing Levels at 4000Hz for four different populations with different noise exposures. Source: Alan Bell, 'Noise', WHO, 1966

A number of percentage-scales of disablement are in use, perhaps the best known of which is that adopted by the aforementioned AAOO. Their scale is based on the average of the pure-tone hearing levels over the 500, 1000 and 2000Hz ranges, which relates to a hearing loss that prevents the subject from fully understanding conversation. But it is clear at the outset that the ability to understand ordinary speech is one firm indication that the noise is not excessive, providing the voice does not have to be raised overmuch, and the speaking distance is a reasonable one. The following table illustrates the relationship between noise levels and the distance that a person would need to keep in order to fully understand normal loud speech.

Table IX: Sound levels for speech intelligibility.
Courtesy: Trade and Technical Press, 1974

Background sound level (dBa)	Maximum distance for intelligibility[1] (metres)
48	7
53	4
58	2.2
63	1.2
68	0.7
73	0.4
77	0.2
82	0.13
87	0.07

[1] For normally loud speech; the distance would need to be roughly doubled if the voice is raised by 5–6dBa.

Many pieces of industrial machinery use an impact to accomplish certain tasks, but the noise they produce is often difficult to measure because it can be accompanied by other more continuous tones. An impulse signal is defined as being very limited in duration and relatively high in pitch. In fact, it is the very shortest acoustic signal that is caused through a collision process sufficiently paced so as not to appear continuous. The British Code of Practice makes reference to impulse noise, although it does not describe in detail the methods of deducting equivalent continuous noise levels. However, the Leq system is said to be able to handle impulse noise more directly, whereas the majority of overseas specifications have introduced special rules and measures which have been described by the Industrial Injuries Advisory Council as making for 'great difficulties' in marginal cases.[16]

It is impossible to measure impact noise. The spectral characteristics of continuous noise can be checked with hand-held instruments to establish band-width limits. But impulsive noise cannot be so readily defined, because a sound level meter will not respond in the time period covered by an impulse. Similarly, the response of the human ear, particularly the inner ear, to impact noise is not well understood. The frequency range is well beyond the range of audibility; the brain does not hear these impulses because of the slow response rate of the nerve and the brain.

The energy levels are high, hence there is no opportunity for any protective reflex mechanism in the ear (as there is when it is faced with continuous noise). It is also not known what portion, if any, of this high frequency energy can pass the outer and middle ear sections, or if it would do damage if it did pass into the inner ear. There is, to complicate matters, some reason to question whether hearing damage from high energy industrial noises can be properly charted at all. It must be remembered that most industrial noise predominates around 250–500Hz, although hearing damage always manifests itself around 4000Hz. The spread of individual susceptibilities is very wide, and is related to the type of noise prevailing. Furthermore, industrial noises contain impulses with a dominant spectral component around 3000 to 5000Hz levels. Burns and Robinson confessed that their study[17] dealt only with continuous noise, and that the problem of impact noise would have to receive more thought later.

7
NOISE AND DISABLEMENT

The reader has no doubt noticed that much of the technical discussion of noise and hearing has been placed within an industrial setting. This, of course, is because the factory or workshop employee is so likely to suffer seriously as a result of noise pollution; and most of the scientific research has naturally been focused upon the industrial scene. Research has also been done on a few other categories of heavily exposed individuals, namely airport workers, military personnel who use firearms, and pop musicians; however, as we saw in Chapter 1, evidence about the hearing loss of the latter group is rather sparse and anecdotal.

What, then, are the least controversial findings that are generally accepted by most experts in regards to noise and its physiological effects? An attempt was made to grade the severity of workers' symptoms in a study by Atherley and Noble,[1] who examined men with twenty years' experience in noisy industries. They reported that the most severely affected had difficulty in conversation, and in hearing a television broadcast. The first few years of exposure saw the onset of high-tone loss established. From then on, it was reported that the hearing of these men declined at a rate that was largely determined by aging, which implied that deafness to some extent is a self-limiting condition. This is why, Atherley and Noble believe, absolute deafness is seldom if ever caused by chronic noise exposure.

However, it is not easy to assess hearing loss when it is a gradually worsening complaint so that the sufferer himself becomes accustomed to it.[2] A phenomenon known as loudness recruitment may be present in cases of hearing loss involving the cochlea. Things sound worse to an affected person because of the exaggerated effects of the masking of adjacent tones.[3] Furthermore, as much as 40% bilateral impairment may be present without the individual's knowledge. Sometimes there is an awareness of a loss in one ear, although audiometry shows both ears to be affected. Words containing many consonants, the frequencies of which may run as high as 10,000Hz, may be hard to catch.

An aural situation known as Temporary Threshold Shift (TTS) refers to any loss from which the ear recovers, however long this takes. An understanding of this condition, sometimes known as auditory fatigue, is essential when considering damage-risk criteria and workmen's compensation. For any person, the extent of TTS is a more or less consistently repeatable phenomenon.

In a typical series of audiometric tests, the hearing of a group of young weavers was measured on a Monday morning before starting work, and a further test performed after they had finished showed that there had been a loss of hearing.[4] Tests on the following morning showed that all the loss had been recovered. Another group was measured twice; on a Monday morning following a weekend, and again after 16 days absence from work. The latter test showed quite clearly that, the longer the time spent out of the noisy environment, the more likely any slight residual impairment would be to disappear completely.

What are the limits that the experts consider ought to apply to factories? In the UK, the Department of Employment thinks that 90dBa ought to be the maximum, in the light of present audiometric knowledge and of the feasibility of getting noise levels any lower. This is supported by the Confederation of British Industry (CBI), but is opposed by the TUC which wants at least an 85dBa limit.

In the USA, the EPA have said they favour the 85dBa limit, with the long-term aim of reducing it to 75dBa, with a 3dB time-intensity trade off; i.e., an increase of 3dBa would be allowed for every halving of exposure. The EPA's recommendations are in line with the stricter British standards as formulated under the equal energy rules. The US occupational noise laws are known as the Walsh-Healey (W/H) rules: they are more lenient (from the point of view of the industrialist) as they allow an increase of 5dB for every halving of exposure. In 1974 the EPA recommended that the W/H rules be brought more into line with the European Leq system. The W/H rules themselves are based for the most part on the Threshold Limit Values, formulated in 1968, by the American Conference of Governmental Industrial Hygienists (ACGIH); their recommendations are given in Table X.

Some authorities believe that pre-exposure to 78dB can modify the development of TTS in reaction to more intense sounds.[5] Within limits, the amount of shift produced by a noise of given intensity is greater for high frequencies than for low. Further, so long as the TTS two minutes after cessation of the fatiguing stimulus is less than 50dB, the rate of recovery on a logarithmic time-scale is proportional to the initial shift, and recovery will be complete about 16 hours after a 2-hour exposure.

Noise induced permanent threshold shift (NIPTS) refers to where the shift is no longer temporary, and is the official definition of sociocusis. The EPA considered that for each 10dB of hearing loss, sound energy would have to be raised by a factor of ten for a particular sound to be heard.[6] To test the impact of NIPTS, Gallow and Glorig performed an audiometric study on 400 men aged between 18 and 65. They confirmed that exposure to high-level industrial plant noise (102dB overall in the spl octave bands spanning 150 to 9600Hz) caused the hearing threshold to rise rapidly (i.e., caused the level of hearing acuity to diminish) over the first 15 years of exposure, after which it levelled off in the higher frequencies of 3, 4 and 6kHz. By contrast, with hearing levels at 500Hz, 1 and 2kHz rose more slowly over exposures of some 40 years.[7]

In their study, on the question of noise exposure and NIPTS, Burns and Robinson screened out workers with otological abnormalities, and allowed time to eliminate the TTS. They then measured differences among people exposed to varying amounts of noise in their working life, quantifying the total noise exposure of each individual, using the equal energy principle. Another well known test carried out in the USA by Baughn did not include otological screenings, making the test more representative of industrial workers.[8]

Arguments abound concerning the merits of epidemiological studies used to predict relationships between noise exposure and hearing impairment. The EPA uses three sets of data (from studies by Baughn, Burns and Robinson, and Passchier-Vermeer) in its analyses, which it considers to be all approximately the same. The American Occupational Safety and Health Administration (OSHA), in its response to the EPA's criticism of its more lenient noise standard of 90dBa, rejected all other findings in favour of that of Burns and Robinson; although Dr Burns has himself maintained that OSHA has incorrectly applied his formula in a way that seriously understates the hearing damage to be expected under the 90dBa standard.

The EPA believe that OSHA's analysis of NIPTS, based on the hearing risk for averaged frequencies of 500, 1000, and 2000Hz is inadequate. They say it fails to account for hearing loss in the critical frequencies above 2000Hz. The EPA instead analysed hearing losses with more attention to the higher frequencies. They asserted that while damage to hearing at the 4000Hz level occurs mostly during the first ten years, other critical frequencies, such as 2000Hz and 3000Hz continue to be affected as exposure is prolonged.[9]

The figures in Table X should be compared with the marginally stricter recommendations advanced by the British Occupational Hygiene Society (BOHS). Their criterion of 1971 is based on the total noise intensity and provides for a period of 8 hours daily, 5 days a week for a duration of 30 years.

The ACGIH and the BOHS criteria both base their figures on the total noise level defined in dBa, although the BOHS curve is more restrictive; i.e., it allows a shorter time in an environment the noise-level of which is identical to that of the

ACGIH criterion. Valcic claims that it is very difficult to make comparisons between the two sets of figures, since they do not relate exactly with the elements of noise to be measured.[10] This lack of comparability also applies to the British Association of Otolaryngologists' (BAO) suggestion that the level of hearing loss appropriate to the requirements of the Industrial Injuries Act would be an *average* of 40dB over the 1000, 2000, and 3000Hz frequencies.[11]

Table X: ACGIH recommended noise limits.
Source: ILO, Geneva, 1974

Exposure time (per day, in hours)	Noise (in dBa)
16	80
8	85
4	90
2	95
1	100
½	105
¼	110
⅛	115[1]

[1] It is recommended that exposure to continuous or intermittent noise louder than 115dBa should not be permitted. Exposure to impulsive or impact noise should not exceed 140dB.

Table XI: BOHS recommended noise limits.
Source: ILO, Geneva, 1974

Exposure time (per day, in hours)	Maximum noise level (in dBa)
8	90
6	91
5	92
3	94
2	96
1	99
0.5	100

We saw above that there was some disagreement between the EPA and OSHA about which has the correct reading of permanent threshold shift. Once again, the question of *feasibility* comes to the fore; OSHA says that, as there is limited data relating to technological possibilities of meeting the 85dBa limit, the current 90 Walsh-Healey limit should be retained. The EPA, in reply, cite the Bolt, Beranek and Newman report, which concluded that the lower limit can be achieved using existing technology. The costs to industry, although appearing to be high, did not, in the EPA's view, reflect the various economic alternatives to the industry-wide threshold limit approach to the setting of standards. At the root of the controversial disagreement between OSHA and EPA is the exhibition of wide variations in industry of different levels of noise exposure, which naturally leads to corresponding differences in the cost of complying with standards.

The issue of employee-efficiency is not of interest only to public policy makers and employers. The important question of well-being connected to that of efficiency is one that every gainfully employed person is aware of. Even so, issues involving efficiency are conceptually distinct from those concerning health, although experts often gratuitously relate the two. Cyril Duerdon, for example, asks several questions to determine whether a hazard exists, among which is the question of whether workers find it difficult to speak.[12] (Apparently between 15 and 20% of all shop floor employees work in surroundings that make it impossible to talk in normal conversational tones.[13]) Difficulty in verbal communication is not medically related to health; neither is it necessarily related to overall factory efficiency, or to individual performance. The inability to talk may be to the advantage of production since it may obviate crucial moments of inattention.

So, one must consider whether noise is distracting (and hence impairs performance); whether it it has no effect; or whether, as Kryter suggests,[14] it is beneficial and attention-sustaining. What is clear is that the influence of noise must be shown to be causally distinct from the influences of other socio-economic and cultural factors. This calls for the establishment of carefully selected control groups.

The problem in the past has been the lack of sufficient comparability in work situations that on the face of it have similar incidences of ill health and accidents, but which have substantially different noise levels. Experimental comparisons can of course be useful in building up comprehensive theories. In other words audiometric instruments can measure both the objective noise and its subjective impact, so that at least a few general principles about the effects of noise can be advanced, even if they do not hold true universally. For this reason some of the well known experiments that have been performed on workers in noisy surroundings tend to be cited as established evidence.

For example, an experiment by Corcoran showed that sleep loss merely compounds the decline in efficiency caused by noise.[15] Yet, in a noisy atmosphere, the performance picks up before declining again, so that the average performance is good (in effect, noise can cancel out the effect of sleeplessness, especially in serial reaction tasks). A state of arousal can even be stimulated aurally.[16] And because the effect of noise on the worker is so subjective, the opinions of the acoustician are rather inconclusive. The subliminal features of noise must be carefully distinguished from that which is consciously annoying. Pleasant background music – or even a moderate steady pitch of sound – may both stimulate worker performance and aid concentration.

In the past, there have been two types of experiment of worker vigilance. The first was the dial scanning test: the subject had to monitor a series of dials in tests numbering from three to twenty. Second were the serial reaction tests: a series of coloured lights appeared, and when a corresponding coloured button was pressed the display changed. In the former test there are two reports of no impairment and two of a perceived decline in performance.[17] In the latter the most significant study, carried out by Broadbent, showed a fluctuating performance initially but, as time passed, the number of operator errors rose appreciably.[18]

In a study done by Weston and Adams in the 1930s in a textile weaving plant, a group of weavers were asked to wear earplugs. They did so reluctantly since they did not share the belief of the management that the noise would be greatly lessened or that earplugs would make them more efficient. This is thought to be a significant factor, because the improvement in their work was not due to psychological factors such as might usually be attributable to the greater motivation of workers in a test. A Broadbent and Little study (1960) showed that a diminution of 10dB led to an improved rate of working on film-perforating machines (although the same increase occurred in a control group). Another study demonstrated that an acoustically treated room had a rate of machine noise of only one fifth that of a controlled untreated room.[19]

Psychologists have tried to make experimental situations more precise by eliminating variables in field experiments. Broadbent and Mackworth showed that by using continuous broad band noise with a fairly flat spectrum at levels up to about 100dB overall (i.e., with an even background noise), efficiency declined at amounts beyond 90dB on a subjective scale of noise applied to motor vehicles, and would be between 'noisy' and 'excessively noisy' to the extent that conversation became difficult.[20]

Thus, most studies of this kind hint that the deleterious effects of noise do not affect the speed of performance, but rather the accuracy. Sudden changes in levels either up or down can result in momentary disturbances,[21] such as wrong button pushing and inattention. It is also worth pointing out that the introduction of unfamiliar noises has only a transitory effect; the decline in performance is noted, and so corrected.

Precision-made instruments are now available to measure either human hearing or the general ambience level of a particular area. The general concept behind the process is similar in all cases. Knowledge about the distribution of sound waves, with time, enables the scientist to formulate techniques whereby *precise* measurements can be made of the fidelity and range of an individual's own level of hearing with an audiometer. Acoustic data is converted into electrical signals which may be amplified and measured on the indicating dial, and calibrated directly in decibels. Some meters are capable of correctly displaying the peak amplitude of even short-duration impulses. The Pure Tone Manual Audiometer emits a continuous tone, and is controlled personally by the examiner after fitting the subject with an earphone. William Burns maintains that the zero corresponds to the normal threshold, and the hearing level value is defined as the deviation in decibels of an individual's threshold from the normal threshold value to which the audiometer is adjusted.[22]

A-weighted sound is usually used in audiometers. In other words, as briefly discussed earlier, a selective filter is built into the device to make it less sensitive to high and low frequencies. Several 'dose' meters – some of which are pocket-sized – are made to help assess continuous industrial noise on the Leq system. Many conform to the Department of the Environment Code of Practice, and measure 100% for an exposure to 90dBa Leq for a period of 8 hours. Care has to be taken with some imported meters, as the USA has assigned a value of 5dB to represent a doubling of energy, whilst in Europe 3dB is thought to be more suitable. In a self-recording audiometer, the subject himself presses a switch when he hears a tone which gradually decreases in intensity. He then releases a switch when the tone is no longer heard and the information is recorded on a card inside the device.

Various portable noise-meters are also commercially available for the measurement of impulse noise, but their usefulness will be of a limited nature until a more standardised method becomes available for assessing hearing loss from their readings. Precision-grade meters are rather more complex, combining the features of the survey meters and those used to test individual hearing (often known as industrial grade meters). However, they have additional circuitry to enable the determination of many types of acoustic signals. They rely on a stable and linear microphone to convert acoustic data into electrical signals which may be amplified and measured on the indicating meter, and is the only grade meter that meets all current British standards.

These last three chapters have explained in some detail the relationship between acoustic energy and human health. Armed with this knowledge we will, in the next two chapters, look at how these harmful emissions, from all the major sources, can be physically controlled.

Part Three

THE PHYSICAL CONTROL OF NOISE

8

DESIGNING OUT NOISE

One of the foremost difficulties in abating noise arises from the fixed fabric of cities from which noise emanates. Roads, houses, factories and airports are already with us, and they have a long life expectancy. So the urban planner must play a more important and strategic role than the acoustician in the *long-term* attack on noise.

This chapter and the next will be discussing the panoply of technological and engineering measures that can be adopted to control noise from machines and appliances. But site-planning is a more comprehensive approach; it is where abatement in effect takes the form of isolating the noise transmitter and receiver at the initial stages. Problems exist that are by their nature intractable without the help of the professionals working as a team. And it is not easy to predict when a particular process will involve excessive noise. It is even more difficult to decide on the correct measures to be taken; they could, for instance, entail simply restricting night work or fitting insulation of some sort.

Nevertheless, there is an essential almost idealistic logic to site-planning, whereby buildings are used to shield other buildings from noise, as has been done at Thamesmead in south London and one or two other places, where offices, shops and garages are sited in between the main industrial and residential areas. Similarly all noisy processes can be located in one building, as this is cheaper than insulating scattered buildings, and one noisy process may mask less noisy processes. In all this, the planner has to visualise the range of daily commercial activities, so that as a result lorries in and out of factory yards can be given a clear run without the need to manoeuvre behind a perimeter wall, etc.

The immediate responsibilities for noise control clearly lie with architects, designers and engineers. Berry and Horton have pointed not only to the duty of the scientist but to that of the doctor and urban systems-analyst in considering the issues involved.[1] A lack of knowledge and foresight is an obstacle in assessing how well abatement policies and strategies are going to work. The failure to plan ahead can culminate in the wasteful stripping down of both buildings and machines and the fitting in of sound absorbing devices (i.e., retrofitting) after it is realised how much of a nuisance the noise problem is.

It is now increasingly the practice for architects to consult with acousticians at the design stage. The acoustician, after conducting preliminary surveys with the aid of extra information provided by the heating and ventilating engineers, can draw up noise criteria curves (NC) (based on the concept of optimum reverberation times plotted against the volume of a room according to the proposed usage) for all areas of a building to provide the data from which the other experts should work.[2]

A growing number of manufacturers are now providing noise spectra information. This greatly helps the predictive powers of the acoustic technician when drawing up his NC, as he must consider the type and quantity of appliances in industrial and residential buildings, including servicing aids such as ventilation-ducts, refuse-disposal chutes and power-generators. However, little foreseeable change is likely in the technological suppression of noise unless designers of machines are legally required to consider the effects of new artifacts on the environment. Of course, such a requirement would put a great strain on the designer's creative skills. He must attempt to design noise out of a product, but keep costs as low as possible while at the same time trying to avoid reducing the appliance's performance. Designers, it must be remembered, are not radical innovators; they tend to think in terms of traditional machine components which are perennially noisy, and any mooted changes are likely to be gradual and cautious.[3]

Nevertheless, one of the most valuable contributions the acoustician can make is to give an exact assessment of the right compromise between effectiveness and economy. Under-silencing is just as wasteful as over-silencing. Noise control as an afterthought, when everything has been built and installed, is invariably more costly than having the correct treatment applied in the first place. The management of an industrial plant might well consider that *post-situ* remedial work is just too expensive, and might prefer to manoeuvre and to negotiate with private nuisance claimants seeking damages, rather than to make modifications to the plant. Indeed, the exigencies of fulfilling business orders may require *noisier* machinery to be installed, plus the commencement of night work, to get a back-log of work down. The employer may take advantage of weak or piecemeal legislation, rather than innovate. In view of the lack of specific occupational noise legislation in Britain, the Factory Inspectorate cannot prosecute the vast number of industries that fall beyond the purview of the three existing statutes, or under local health regulations.

Most occupiers of flats in large apartment blocks suffer from noise transmitted through adjoining walls. The transmission of sound from one room to another can be both direct and indirect, and the relativity of the two is important in the design of a satisfactory insulation system. But increasing the mass of walls and partitions has little practical effect, unless they have some kind of resilient support or isolation at the connecting surfaces. 22cm(9in) – brick plastered walls can reduce noise by up to 50dB, but properly constituted double walls and windows (where the second pane in a double-glazed window is at least 30cm (1ft) apart from the first, and is independently mounted in the window frame) give more insulation for a given total weight, although they may cost as much as heavier single-walled constructions.

All acoustic ceilings and barriers with holes in them perform the principal building function of absorbing sound. They are generally of a fibrous and porous nature, as they work on the principle that hard flat surfaces reflect sound in the same way that a mirror reflects light. They can come in a variety of forms: tiles, wallboards, acoustic 'spray-on' plaster, acoustic blankets, etc. Most fibrous screens or blankets can absorb between 60 and 80% of sound. But frequency must be taken into account, as most absorbers operate best at medium frequencies.[4] Unfortunately transport and industrial noise predominates in the lower frequencies and the effect of insulation treatments is to make the sound spectra distorted even more towards the low frequencies, so that the house occupant often feels rather than hears sounds.[5]

Another form of directly transmitted sound is impact noise (from footsteps, door-slamming, etc.) which comes through ceilings. To overcome this the architect can design discontinuous or 'floating' floors on a mineral wool quilt[6] or screed. This concept of isolating noise-generating materials from solid floor surfaces is particularly important in industrial premises. It is therefore essential that the architect be aware of the purpose for which the building is being used. What must be avoided is structure-borne sound arising from machines bolted to ordinary screed concrete surfaces, in buildings with single glazing that have roofs of asbestos sheeting and walls of 22cm (9in) brick. Even without the architect, resilient mountings or damping elements could of course be installed in advance to act as a buffer between a heavy component and the floor foundations. All connexions to a generator engine or alternator (such as fuel pipes, conduits and air ducts) must be flexible, as rigid piping fixed to walls can transmit noise down solid electrical leads.

Of more concern to the architect is the resilient mounting between the room foundations and the surrounding environment. It may be possible to mount heavy equipment on an isolating raft consisting of a concrete platform about 15cm (6in) thick, placed on top of a structural floor but isolated from direct contact with it by a layer of compliant material such as the mineral wool quilt mentioned above. A number of these are commercially available to varying densities according to the load and forcing requirements of the plant manager (this can be worked out from the known weight of the machinery and its rpm). With mixed light and heavy equipment, the architect can design an anti-vibration floor where the structural surface can protrude through the raft at points where heavy (and suitably mounted) machines are installed. The principle can work in reverse; a concrete floor can have occasional pits sunk into it into which machines can be placed and which can be surrounded by a moat-like layer of compliant material.

The *ab initio* approach to the containment of traffic noise means that purely engineering modifications are of a secondary nature. Certain landscaping changes can be made, such as building roads underground. However, we need to reiterate that cost factors have a decisive bearing on the matter. As the Road Research Laboratory point out, a road tunnel is thirteen times more expensive than a road[7] so an underground carriageway is generally built rather as an entirely new project than as a replacement for an existing roadway. Planners in the future will no doubt be considering as a matter of routine the sinking of motorways and skirting roads into purpose-built valleys with vertical or sloping banks. Full advantage can be taken of contouring to ensure that as many barriers as possible are placed between the planned carriageway and nearby houses. Asphalting over concrete can possibly reduce noise levels by 5dB.

Cars are already reasonably quiet considering their power and speed. No really great reductions in present noise levels can be obtained from modifications to the engine; to all practical purposes, only a fundamental redesign of the internal combustion engine, or a radical change in the motive power (say, to turbine, jet, gas, nuclear fission or battery driven engines) will bring about really *noticeable* emission reductions.

There are several important sources of car-engine noise: intake and exhaust, gearbox, cooling fan, and intake noise from a high speed airflow through valve seats. According to Bugliarello *et al.*, one way to reduce engine noise is to increase the number of cylinders and reduce the size of the engine bore.[8] As it is the initial pressure rise in the cylinder that determines the level of combustible noise this pressure can be reduced whilst the engine power is brought up again by pressure charging. At present, some British manufacturers are replacing the OHV with the quieter OHC engine.

A reduction of some 6 to 10dB is possible by completely enclosing the engine casing with absorbent lining, but this suffers from the disadvantage of poor engine cooling, increases in weight and poor accessibility. To be able to effectively enclose an engine without seriously overheating it would entail major changes in configuration. Suitable silencers can reduce the vibration and noise of the intake valve, and exhaust cams can be redesigned. The remaining crankcase and cylinder block noise can only be substantially contained by greatly increasing the stiffness of the engine walls, by adding on layers of damping lightweight metals such as magnesium, to avoid increasing the weight. As we have seen from Chapter 2, the turbo-jet is gradually being replaced by the quieter high-bypass turbofan jet which was designed to comply with stricter official regulations. The retrofitting of turbojets often takes the form of redesigning the exhaust ducts or nozzles. This can have the effect of shifting noise higher up the noise spectrum, so it can be more easily absorbed into the atmosphere. The high-frequency whine of the fan-blade can be minimised by acoustically optimising the spacing of the vanes and blades.

The NAC make reference to a special 'hush-kit', which includes the fitting of sound absorbing panels in the air intakes to reduce noise radiated forward from the compressor. These panels can also be fitted into the bypass ducting and the jet pipe to minimise noise from the bypass fan and the turbine respectively.

More radical changes in design would entail actually converting a low-bypass engine into a high-bypass one by fitting a new large-diameter fan and by mounting the engine in a new nacell. But the big changes in the structure of the centre engine position would bring about ground clearance problems where engines are under the wings.

Of course, exhaust velocity can be simply decreased by reducing power after take-off, or by restricting the payload or fuel so as to permit reduced engine power. Planes could also approach to land at an airport at a steeper angle than usual, say six degrees instead of the current three degrees. This would mean that at 19km (12 miles) from Heathrow, aircraft would be flying at 1500m (5000ft) instead of 750 (2500ft), and would be emitting 63dB instead of 78.[9] Some progress is being made with such mooted changes in flight procedures; it is now common practice for jets flying over dense urban areas to throttle back the engine once a safe height has been reached.

9

CONTROLLING INDUSTRIAL NOISE

Although the impact of industrial noise on the community (in terms of the number of people at risk of losing their hearing) is the most serious, the planners and managers of British industry have had very little success in abating the noise emissions at source. The Factory Inspectorate have a statutory interest in seeing that the most effective, and necessarily practicable, noise control methods are applied, but are often unimpressed by the unsuccessful attempts by plant managers to cope with problems not regarded as exceptionally difficult.[1]

The inspectors observed a frequent failure of engineers to understand and apply the 'most basic principles of noise control engineering'. They noted the inability of technicians to ensure that the work was properly executed, and concluded that there existed a felt need for more engineers to work on noise abatement technology, especially at technician and in-plant levels. The failure of noise control was also seen to arise from the devolution of responsibility to inadequately trained junior staff, who lacked the financial and managerial backing needed to implement major changes in working environments.[2]

Yet there are a considerable number of strategies the factory manager can adopt without entailing his company in a great deal of expense or dislocation. And of course individual machines can be tampered with far more easily and openly than can cars or aircraft, as to modify the latter *in situ* is virtually impossible without major redesigning, which is the manufacturer's responsibility. Let us look briefly at how noise-reduction changes can be most easily brought about.

As mentioned earlier, the location of buildings themselves

forms an important part of the abatement strategy. In an industrial establishment divided into manufacturing and clerical sectors, it is logical to face the manufacturing areas towards any noisy outside sources, such as road and air traffic. Similarly, sectors requiring quietness should be located on the side away from main noisy thoroughfares.

Within this perspective, we ought to distinguish between community noise abatement, and the more industrially specific occupational noise control. The only obtrusive noise arising from a great many factories arises perhaps from outlet exhausts or from ventilation systems. Shipbuilding yards generally face the water, and land is shielded by administrative buildings. In many cases there is no need to make any changes at all. A logical argument can be made against the expense of controlling noise from a loud machine which is seldom approached by an employee.

Inside work areas, the distance between the sources of noise and other working areas can be increased before considering investment in isolating barriers. Automatic machinery not needing constant attention can be placed in storage areas or in an outbuilding. Staff can be rotated from job to job (if this is feasible) when a particularly noisy process is in operation.

Acoustic isolation and independent enclosures are the logical next steps. A solid box enclosure for a series of tumblers used to separate components was claimed to reduce a level of 111dB at a distance of 1.5m (5ft) to 88dB.[3] Many continuous processes, controls and monitoring instruments can be installed in a central console, which can be enclosed in an attenuating refuge – similar to a studio – from which the plant manufacturing process can be observed.[4] If the plant is known to have vulnerable fire hazard areas, a smoking zone can be added, as can a refectory and locker facilities.

It is not easy to lay down comprehensive anti-noise formulae for abating all types of machinery regardless of purpose and setting (and assuming economic feasibility was not an important criteria). For this reason attempts to abate machinery noise at source often begin with analytical forms of measurement.

Appraisals are made by the acoustician about the balance of advantages in terms of the machine's capacity in relation to its weight and size, and handling properties. The transmission path of vibrating and resonant components have to be traced. The individual approach to noise control was exemplified by the West German Engineers' Association (VDI) through their involvement with a noise nuisance arising from iron and steel plants in the Ruhr and Rhine areas. Electric arc furnace emissions could only be reduced with the aid of unique soundproof furnace doors, and by building a second bay around it for shielding purposes, together with mufflers on the ventilating and exhaust systems.[5] But all of these structures and devices could not be used for any noisy machinery. And the people who built the arc furnaces would be the only people who could conceivably understand how the noise emissions could be controlled. Manufacturers of heavy equipment, because of their specialist knowledge of design, performance and manufacture, are often the only available source of knowledge about abatement.

Hence, a question the designer has to ask himself is how specific is the noise in relation to the function of the machine? The modification of machines to effect the method by which their main function may still be achieved offers a further opportunity for noise control. For example, punches in die-sets may be redesigned so as to convert one major impact into a series of lesser impacts, a chopping process can be substituted for a grinding one, or the force of an impact can be lessened and its frequency increased. Some habitually over-cooled electric motors could have the 'fan note' reduced simply by cutting back on the amount of cooling air used.

The technique of changing the process, rather than the function, can be carried further. Pneumatic riveting can be replaced with hot squeezing under hydraulic pressure, rubber buckets can replace galvanised ones and other items traditionally made of steel, such as chutes, gears, rollers and bearings can all be substituted for rubber, fibre or plastic components. Metals can be cleaned chemically instead of being subject to the more noisy high-speed polishing (chipping inside a tank may reach 126dB, but flame gouging will reduce this by 20dB or more).[6] Mechanical, rather than air, ejectors can be used on presses, and the fit between moving parts can be improved. The resonance from metal plates and parts can be reduced by stiffening (e.g., by adding ribs) to increase the inertia against vibration, or by improving the damping with one of the compounds mentioned earlier. For industrial use, 'sound deadened steel' is available, consisting of two sheets of metal bonded together with a visco-elastic adhesive. Components making contact, such as cams, push-rods and driving quadrants can be bushed with rubber or some other compliant material.[7]

An important concept in terms of engineering controls is that of proper upkeep and repair. Rattles can be eliminated by securing a loose panel; squeaking brushes can be quietened by applying a wax candle commutator-surface. If gear teeth are initially incorrectly formed, or have received considerable wear, they will increase the noise level. It is possible to effect improvements by examining the teeth and by filing or scraping high spots, and by checking the depth of the engagement. Vibration from machinery with rotating parts can be reduced by attention to proper balancing. Friction from the cutting action of tools and saws can be contained if the tools are kept sharp, and other friction can of course be reduced by adequate lubrication.

Let us look for a moment at one of the noisiest processes in industry, pneumatic compression. A firm specialising in acoustically modifying pneumatic equipment, Atlas Copco Ltd, has naturally given a great deal of thought to the matter. They discovered that there are three factors in the emission of noise from a pneumatic drill: (1) low-frequency exhaust sound arising from spent compressed air; (2) high-frequency noise coming from the moving parts; and (3) noise from the vibrating chisel hitting the ground.

Copco adopted the method of surrounding the mechanism with a plastic muffler hood which contained two exhaust pipes tuned to the impact frequency of the chisel. This resulted in a reduction of the low-frequency exhaust noise by more than

15dB, as well as much of the high-frequency noise of the working parts. For a pneumatic compressor, Atlas Copco designed a silencer to combat intake noise, which consisted of a chamber mounted between the intake nozzle of the compressor and the intake drum. It communicates with the outside air via a narrow inlet tube, where the low-frequency intake noise is effectively suppressed with little loss of intake capacity. An older compressor can be silenced by being screened off or covered with a sound-insulating cowling.[8]

There are other techniques to reduce compressor air noise. A US manufacturer ensured that his compressors (1) had good quality gears; (2) were made of cast iron instead of fabricated steel (which had better damping qualities); (3) had radiated surfaces that were minimised. He also ensured that specifications to vendors for the supply of components were made more rigorous.[9] Another manufacturer developed a skid-mounted machine that dispensed with the need for a foundation, and this managed to generate 50–75% less noise than conventional reciprocating centrifugal pumps.

Finally, it should be remembered that there is an inevitable limit to the techniques of modifying machines at source. Many machines may be amenable to enclosure or other local treatment, but the problem of ventilation, cooling and personal access should not be overlooked. The acoustic consultants Bolt, Beranek and Newman (for their report 3246 submitted to the EPA) surveyed 68 major industrial plants in the USA and discovered a number of industrial operations that were not susceptible to either engineering or administrative controls. They include certain riveting operations, such as those used in aircraft building; chemical milling operations involving rubber mills; grinding operations in structural steel processing works; and specialised operations involving the use of oxygen lances, rock drills, weaving looms, spot-welding equipment and paint spray guns.[10]

Some industries are regarded as irremedially noisy: examples not included above are those entailing drop forging, hammering and fettling in steelworks, and rotatory whining processes used in shipbuilding, textiles and car assembling. Only a revolution in technology will bring about change, or a decline in the prevalence of such industries. Where emission levels can be reduced through technical refinements at the design and production stage the problem becomes one of the choice of machine-tool. The onus rests with the manufacturers providing other industries with machinery, and they often decline to make available quieter devices because of the higher prices involved.

This is not to argue that extensive redesigning does not take place. New designs arise from innovative methods of research and development. At the redesigning stage, large machines can be taken into anechoic chambers and the effects of modification in design noted. Electric motors can be run at various speeds and flux densities, and the effects of skewed or semi-closed slots recorded. Comparative studies can be made of spur and helical gears that are different in terms of materials, the geometry of the teeth and the viscosity of the oil. The noise of ball and roller bearings can be studied while varying the roundness of the balls, the harness of the races and the finish of the rolling surfaces.[11] A company that managed to do considerable research and development in valve noise reduction was able to provide a variety of silencers for customers, and it developed several noise source treatments such as the 'whisper trim', a specially designed body trim as an accessory to a standard valve.[12] The rate of return on investment on research can be surprisingly good. A US firm achieved considerable reductions in the rate of structure-borne vibration of diesel pumps at a cost of only $20,000.[13]

Earplugs and muffs of various types are widely used in industry, and often represent the only attempt by industrialists to come to terms with the problem of noise. In many ways, the use of ear-defenders is a logical and necessary step to at least minimise the impact of noise on the worker's ear. But as a *normalised* practice it is considered by the Factory Inspectorate to be a 'less attractive' solution than abatement-at-source. Indeed, it can be argued that the provision of such protectors cannot be classed

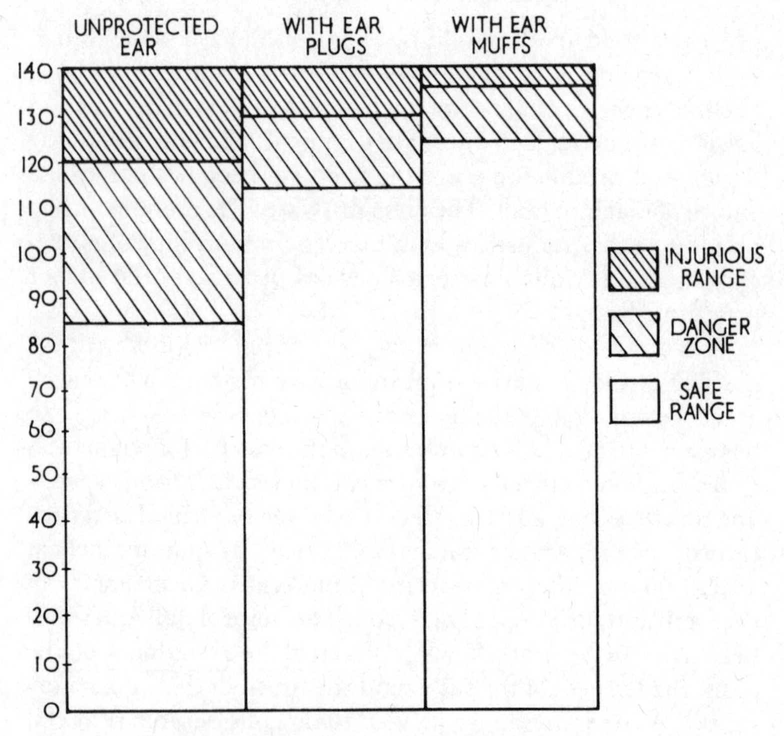

Figure 8 General extent of protection afforded with ear protection devices. Courtesy: Atlas Copco Ltd, 1975

as noise abatement in its accepted sense. But where noise reduction was difficult to accomplish, it would remain an essential tactic, and would provide protection for a greater number of people who might otherwise be noise-deafened. However, they are often thought by acousticians like A. M. Martin to only be a temporary measure.[14] Indeed, under present British law, it is not enough only to supply hearing protection without attempting to reduce noise hazards. And the government decrees that where ear-protectors are supplied they should be efficient, and the employer must assume the responsibility, under the Health and Safety at Work Act of 1974, both to educate employees about how to use them, and to ensure that they *do* use them. The Occupational Safety and Health Act in the USA has almost identical requirements.

What sort of hearing-protectors are commercially available, and exactly how effective are they? There are four main types of device: plugs, semi-insert plugs, muffs and helmets. Plugs themselves can either be individually moulded or disposable. The former are usually made from a form of silicone rubber, and they have a psychological advantage in that it is easier to persuade workers to wear them. The latter type are malleable, and can be made from a variety of substances, such as waxed cotton wool, glass-wool and mixtures of these. Both types are non-porous, reasonably comfortable to wear, and are capable of providing good attenuation values. Ear-muffs are held against the sides of the head by spring-loaded adjustable bands, and give excellent attenuation with the aid of soft

Table XII: Typical mean attenuation and standard deviation characteristics, in dB, of different types of hearing protection. Courtesy: A. M. Martin, University of Southampton, 1976

Test Frequency Hz	125	250	500	1000	2000	4000	8000
Dry cotton wool plugs	2	3	4	8	12	12	9
(s.d.)	(2)	(2)	(2)	(3)	(6)	(4)	(5)
Waxed cotton wool plugs	6	10	12	16	27	32	26
(s.d.)	(7)	(9)	(9)	(8)	(11)	(9)	(9)
Glass down plugs	7	11	13	17	29	35	31
(s.d.)	(4)	(5)	(4)	(7)	(6)	(7)	(8)
Personalised earmould plugs	15	15	16	17	30	41	28
(s.d.)	(7)	(8)	(5)	(5)	(5)	(5)	(7)
V-51R type plugs	21	21	22	27	32	32	33
(s.d.)	(7)	(9)	(9)	(7)	(5)	(8)	(9)
Foam-seal muffs	8	14	24	35	36	43	31
(s.d.)	(6)	(5)	(6)	(8)	(7)	(8)	(8)
Fluid seal muffs	13	20	33	35	38	47	41
(s.d.)	(6)	(6)	(6)	(6)	(7)	(8)	(8)
Flying helmet	14	17	29	32	48	59	54
(s.d.)	(4)	(5)	(4)	(5)	(7)	(9)	(9)

circumaural cushion seals. Semi-insert plugs are moulded to fit the meatus of the ear, are comfortable, and attractive to workers; but the attenuation is poor owing to the loose fit.

The efficiency of a protector may be assessed by comparing the amount of temporary threshold shift attained on working days when the plug is not used. The average of the measured attenuation over the four octaves 250 to 4000Hz is called the noise-protection figure. Figure 8 gives a general impression of the range of protection afforded by ear-plugs and muffs; Table XII illustrates the maximum protection to be expected from hearing protectors that have different levels of effectiveness in terms of the length of time they are worn. To estimate the protection supplied by a particular hearing device, its attenuation-frequency characteristics must be compared with (and subtracted from) the sound level-frequency characteristics of the noise concerned. This, according to the British Code of Practice, will then provide the sound-level frequency characteristics of the noise present at the ear canal.[15]

For example, if hearing-protectors with an effectiveness level of 30dBa are worn in a noise environment of 115dBa, the sound level at the wearer's ears will be 85dBa. But if the user fails to wear them for a total of only 2% of the working day, i.e. about 10 minutes in 8 hours, he will be exposed to 115dBa for 10 minutes which is equal to an equivalent continuous noise level of about 98dBa for 8 hours.

Insert-type protectors have the advantage of being comfortable in hot environments; they can be worn safely with other protective equipment such as glasses, helmets, etc. The cost is low, although disposables may cost more in the long run. The disadvantages are the lower, and less consistent, degree of protection afforded, the additional time needed for them to be fitted, and their tendency to attract dirt into the ear. Muff-protectors can have the virtue of standardisation and good protection qualities. They have the disadvantage of being uncomfortable in hot environments. They are also incompatible with other gear, and the suspension band forces holding them firmly over the ear may weaken with wear.

Part Four

NOISE AND THE AUTHORITIES

10

NOISE AND THE ADMINISTRATIVE FUNCTION

As stated in Chapter 3, in 90% of cases noise emanates from a variety of *manufactured* equipment. It is appropriate to ask, therefore, whether industry can contain the noise problem if left to itself. Most governments tend to believe that industry cannot, and they do not look too kindly on overly lenient approaches to enforcement policies. The encouragement of quieter technologies requires that standards be gradually raised, so that funds and research energies of large manufacturers can be redirected to solving noise emission problems.

Indeed, there exist some extremely compelling reasons for public authorities to intervene in the matter of noise control. Apart from the enormous technical resources that many big appliance firms have (the 'tapping' of which may only occur under some form of regulative control) anti-noise ordinances can be very suitably applied to places of work. Industrial premises offer limited, measurable spaces where noise sources are easily ascertained, and where the legal control of the premises is in the hands of a few known individuals.[1]

A further compelling argument is the considerable educational value of zoning.[2] The panoply of technical negotiations and working papers on the subject represent a form of one-sided public debate which is dominated by the propaganda of official agencies. Furthermore, of course, the authorities not only consider themselves to be the final arbiters in the matter of deciding issues concerning the health of workers and the well-being of residents, they also believe they must exercise their enforcement powers wherever necessary.

Nevertheless, difficulties arise in the intervention process. For example, how and why are policy-making functions assigned to certain administrative institutions, or to local instead of central authorities? The difficulty with present noise-pollution-control practices is that there is more than one approach for the authorities to adopt. Ordinances appear to occur at random at every governmental level. Bugliarello *et al.* also point to the 'excessively zealous approaches' of environmentalists which have resulted in legislators writing largely unenforceable provisions into anti-noise codes.[3] Having a maze of legislative tools is doubtless, as Harvey and Hallett suggest, associated with the process of industrialisation and urbanisation.[4] Every aspect of the environment is in some way either a subject of the law, or defines it.

In Britain, unlike countries such as France where there is a system of codified statutory law, there is much emphasis on private prosecution as a means of remedying some social disamenity; but as we shall see later there is some evidence that things are changing. Furthermore, British planning and zoning arrangements are curious quasi-statutory hybrids. Local authorities will often have powers given to them under an Act of Parliament to help them site and locate industrial and residential areas, but in exercising their powers they are virtually obliged to collaborate with important members of the community (instead of dictating to them) in order to arrive at reasonable and economical solutions to the problem. A planning decision will also have a different impact on, for example, an industrial manager, than would a clearly enunciated law.

Anti-noise laws, at the municipal, jurisdictionary and local levels, define noise in a number of ways. Older laws forbid 'unnecessary or unreasonable' noise. The Noise Abatement Act of 1960, and similar acts in the USA borrow definitions for statutory nuisance from English common law. In other countries, such as France, there are no specific acts dealing with noise, nor any explicitly worded prohibitions concerning pollution in their codified legal system. The law simply says that pollution must not occur.[5]

Yet Levi and Colyer suggest that, in most countries, the law has failed to prevent noise at source. For example, the imperfection of Belgian legislation arises not from its leniency but from its terminology. A new and noisy installation of some kind will be refused permission if it is likely to cause *abnormal* discomfort to the neighbourhood. But is the noise 'abnormal' when it comes from a new manufacturing plant that has started up on an old-established trading estate? The Belgian rules to protect the health of the work force are also very imprecise. The absence of any uniform criteria merely encourages the local authorities to adopt a flexible approach to attract industries to their areas. Recourse to the civil courts is frequently ineffective; there is difficulty in proving 'default through lack of concern', 'abnormal circumstances', and so forth.

The plethora of pending legislation at State and Federal level in the USA poses rather different problems. One concerns the importance of lay as well as scientific standards (i.e., the subjective versus the objective). Another is the absence of any single method of noise measurement which accurately describes environmental cause and effect upon which proper criteria can be established. So for reasons arising from acoustics and audiometry, the orthodox environmental approaches of what A. F. Meyer refers to as 'primary (health-related) and secondary (welfare-oriented) standards' are not directly amenable to control by the authorities, if this implies strict enforcement and severe penalties.[6]

Many of the theoretical problems of noise abatement concern not the nature of legal ordinances, but evaluative and social issues. It would not be fair (as we have seen from Chapter 1) to say that *all* noise is a public nuisance, and as a result adopt punitive policies towards noise-makers. But who is to decide upon the final scope and shape of the nation's anti-noise laws?

The failure of governments to prevent neighbourhood noise in the past has not been due merely to legal or judicial confusion. All legislative processes are characterised by prominent arbitrative features. The judgements that have to be made about perceived social inconveniences and their impacts have never been easy. They can only become more complex as social and moral issues – from abortion to nuclear reprocessing – become political issues almost overnight. The disamenities of unregulated traffic flows and industrial activity, the problem of an equitable balance, the loss or benefit to society as a whole or to sections of society in particular are all intricate problems of judgement. Fortunately, they are by no means intractable.

The Buchanan report of the 1960s discussed the amounts which can be usefully invested in motor vehicles on the one hand, and in facilities for their use on the other. The ethic of the writers of the report decreed that the limitation of motor traffic in cities was 'distasteful' but unavoidable.[7] It was the social inconvenience of some acts that ought to be contained, although such acts were not necessarily morally wrong. This is an ethic of some considerable importance in practically all environment and public hygiene laws. Yet it is one that places highly complex choices onto a legislative assembly that may not, for institutional and cognitive reasons, be able to handle them.

In the first place the traditional political processes of reconciling sectarian diversity can comfortably be accommodated within the domain of pressure-group politics. Moral issues, even where the process is subject to overwhelming bipartisan lobbying pressures, can generally be resolved by able ministers. But in most environmental matters, the parliaments of capitalist countries are being asked to adjudicate on the trade-off between environment and industry. They are hence having extra political burdens placed on them. The resolution of such problems is difficult largely because of the highly technical and clearly judicial elements involved. The noise-abatement issue highlights several key points about the scope and characteristics of government in this attempted trade-off situation. The first concerns the state management of some economic processes, and the second concerns the operating principles of government.

The speed of technological innovation suggests that pollution problems may be self-resolving in the long run. But unfettered economic growth suggest more than this. A broader argument maintains that as the regulatory approach to economic problems has failed in the USA there is some doubt as to whether it will succeed in the sphere of pollution control. The American Commissions, like those in Britain, tend to acquire a technical bias that can lead to rigidity; vested economic interests tend to gain influence over them. A divergence appears between the pursuit of scientific objectives operating by strict economic criteria, and the pursuit of public policies.

There is also the problem of the divergence of pace. One argument implies that public strategies become ineffectual, since the spasmodic rate at which innovations occur would render a comprehensive scheme unworkable and unjust. It would inevitably be tardy, too, as a result of a long period of investigation and review by public committees. More significantly, the speed of exploitation of a profitable new technology implies that there will be an absence of early control standards until consultations – of a probing, cautious nature – result in an acceptable policy framework for control.

There is a logical argument in opposition to the *laissez-faire* doctrine. Highly urbanised societies become complex and crisis-ridden, and the growth of socialist-oriented policies seems to become inevitable. The greater conspicuousness of social problems necessitates the spending of public money, as the perception of public nuisances grows with rising expectations of well-being. In another sense, as environmental problems are the result of resource use,[8] society has the right to expect resources to be reallocated to counterbalance the disamenities. The implications of this are both subtle and interesting.

The legal notion of the democratic state is that of a trustee of common property resources, rather than a director or user. Yet there always seem to be subtle pressures put upon western governments to act as directors rather than as trustees, and it is a trend that is encouraged by the judiciary. Hitherto, the courts played a critical role in the selection of environmental policy because of their ability to question the constitutionality of legislation.[9] In addition, the courts aim to strike a balance to retain the biggest measure of public use consistent with economic growth. But the difficulty for the courts is that in spite of the importance of an expanding body of case law, they can rule only on specific cases. Significant trade-off issues between environment and industry have again to be referred back to the government. As the government is aware that the framing of welfare legislation in line with only technical criteria may be costly, it probably allows firms to plead in defence that the 'best practicable means' (say, for abating occupational noise) had been used.

This is a highly liberal clause: the noisemaker can maintain that his level of noise-containment is all he can afford, and government officials are not well equipped to evaluate such a defence.[10] As such, with a major exception that we will refer to later, the government adopts a passive rather than a coercive approach to the control of resources. It tends to relieve the polluter of direct involvement, and places the onus on independent academics and technicians to arrive at new and cost-effective methods of control. But it should not be forgotten that the government's desire for noise control stems from considerations of welfare, which involves the question of human values. The experts they would need to employ to monitor aircraft, traffic and industrial noise would very likely be *technically* qualified people who perhaps are not the best fitted to deal with the subliminal or subjective features involved. A technician may be unaware whether the aims of his task is to protect hearing-sensitive minorities, or majorities, or whether he should be advocating the cheapest noise control methods, or the most effective.

Following from the above, we might now be able to see how shortcomings in formulating and administering anti-noise rules actually arise. The two main reasons are: (1) the highly democratic nature of the enforcement process; and (2) the difficulty of arbitrating upon complex issues that are both social and technical. In recognition of these two aspects, a new feature of government has evolved: the specialist agency (like the Noise Advisory Council or the Health and Safety Executive) to which legislative or strong advisory powers are entrusted. It is at this stage that the *laissez-faire* principle of the courts is attempted to be undermined by a more committed, radical stance. The NAC has in fact stated quite clearly where it stands:

> . . . the benefits of the use of factory machinery accrue directly to its owner and indirectly to his employees and to society as a whole; whereas the disbenefits of the noise emitted fall in different degrees upon the workers concerned and upon a very limited section of the general public in the immediate locality. In these circumstances positive action by public authorities is clearly necessary to redress the balance.
>
> A quarter of a century ago this proposition might have seemed radical and questionable. The first cautious and tentative recognition by Parliament of its validity occurred only just over a decade ago. Today it is universally accepted and sounds almost trite. This is a measure of the rate at which both the problem of noise intrusion and public awareness of – and dissatisfaction with – it has grown.[11]

To understand the significance of the setting-up of new agencies, and the declared motives of their spokesmen, we need to look at one or two theoretical models of government for a hint of the kind of power structure under which the leadership operates.

Sir Geoffrey Vickers, in his *The Art of Judgement* divides organisations into 'user-supported' and 'public-supported' concerns; where nationalised industries fall into the former, and governments into the latter. And rather than viewing the purposes of government or organisations as primarily goal-oriented, Vickers substituted the concepts of *optimising* and *balancing*. This meant that decision-making is either profit-oriented – gaining the maximum economic value for a given service consistent with efficiency – or is regulative – where the system contains built-in governors and balances.[12] There is another model, suggested by William B. Lord, which compares a 'bottom up' with a 'top down' approach. The former conceives of regulations beginning at local level and proceeding upwards through agency channels to central government, in the process clearing a number of technological, economic and social hurdles so that the final proposition is neatly reformulated. The 'top-down' concept sees broad goals of public policy decreed by the leadership which succeeds in filtering down the system and ultimately finding the appropriate instruments for the execution of policy.[13] Do any of these models ring true in regard to US or British practice? It is certainly possible to discern changes in the style of administration during the past twenty-five years. UK attitudes towards the quality of the environment have been less striking than those of the USA. Britain has had a longer legislative history in environmental matters, as witnessed by the continuing programmes for the control of water, air and oil pollution, as well as noise abatement. For example, community noise emission control dates in Britain from 1960, and from only 1972 in the USA. Secondly, for nearly thirty years, British planning policy has incorporated some notion of environmental assessment.

However, Britain has tended to avoid the formation of the grand administrative programmes favoured in the USA. The reason for this may have been political and economic weakness. Excessive noise emissions are a menace to rich, rather than poor, countries. It becomes easier to think about industrial waste of all sorts when cost-consciousness is for once no longer of paramount importance. And because of the imperatives of financial stringency in the past, Hufschmidt believes that British policy has avoided the full commitment to anti-pollution measures that have been adopted in the USA, in favour of more balanced, and restricted, programmes.[14]

In 1970, the US federal government decided that an unpolluted environment was as important as a stable economy, thus consolidating the longstanding approaches to environmental issues. The National Environmental Policy Act (NEPA) of 1969, and the newly created EPA were the two chief instruments of this basic restatement of policy. This broad, codifying Act created the Council on Environmental Quality (CEQ) to advise the government and to oversee the compliance of the federal agencies with the NEPA:

> (The CEQ) requires federal agencies to consider the environment, along with traditional economic and technical factors,

> before taking major actions. It asks agencies to examine their laws and regulations to identify requirements that prevent accomplishing environmental objectives and to take actions to correct these deficiencies . . . [15]

This grander, strategic, approach was also spreading to Europe. The EEC Council of Ministers declared that the best environmental policy 'consists in preventing the creation of pollution or nuisance at source', and requested that an evaluation be made of the effects of pollution on the quality of life. Hence, one of the most important concepts sustaining abatement policies, as Duerdon reminds us, is that of prevention being better than cure.[16] In regard to noise, such attitudes have been officially expressed in the British Circular 22/67, issued jointly by the Ministry of Housing and the Welsh Office in 1967. Paragraph II reads:

> The ministers consider that control in the light of development plan provisions has done much and will continue to do much to prevent the establishment of new industry in places where it could cause nuisance by noise.[17]

So in the UK, at about the time of the NEPA in the USA, the recognition grew of the need for more comprehensive policies. The Labour government took the first official step concerning overall environmental quality control by giving the newly established Secretary of State for Local Government and Regional Planning coordinating responsibilities for all anti-pollution measures. Under the new Conservative administration, the coordinative task was assumed under the aegis of the present Department of the Environment.

A prime motive for establishing grand environmental programmes also derives from the need for an ultimately identifiable and responsible authority for anti-pollution measures, in addition to the back-up agencies and advisory bodies. This demand for a 'single referral point' was made by the scientific adviser to the GLC.[18] Centralisation also facilitates the lengthy research and consultative processes that are needed both prior to and after the passing of Acts. Furthermore, it was becoming recognised that certain injustices in the way the common law remedy for nuisances was operating demanded the strengthening of the government's powers of prohibition in regard to certain antisocial acts.

In the 'sixties noise graduated to become a major social problem. The government set up a committee, chaired by Sir Colin Wilson, which published the report on noise in 1960. It was instrumental in passing the Noise Abatement Act of that year and in founding the Noise Abatement Society. From that year onwards a noticeable change has taken place in the public administration of environmental matters. Although the 'bottom-up' process is still characteristically British (and still largely applies in the planning field), it now seems to be changing to the 'top-down' principle. The impact of the environmentalist lobby on Parliament, and the declaratory aims of Ministers to tackle the problem have succeeded in enhancing this image. It is illustrated more concretely in the growth of statutory law (where people can complain to magistrates rather than to the council or county courts) like the 1974 Control of Pollution Act, Part III of which supersedes the 1960 Act and the earlier public health acts.

The USA also seems to be putting 'optimising' principles into practice. The establishment of regulatory commissions seems to be an attempt to respond to specific problems, and to alleviate the hazard of governmental overload. This extra governmental burden arises mainly because, as we have seen, societal decisions previously made by the courts are increasingly having to be made by the political leadership which bases its decisions on consultative, i.e., political, techniques.

There are certain advantages in the agency, or Commission, system. It can develop the necessary expertise, and can attract the right technical staff. It can be seen to be politically and industrially independent; thus it works in the public's interest. It can operate less formally and more flexibly than, say, judicial institutions.

However, in the USA experience has shown that the Agency cannot make itself as independent as it would like. For example, the Walsh-Healey Public Contract Act empowered the Secretary of the US Department of Labor to set standards for health and safety in firms doing more than $10,000's worth of business annually with the government. The limits then were not set above the level at which experts had shown hearing loss might occur. With the establishment of the EPA, and the growth of public and official concern about noise pollution, a recommended ceiling of 85dBa was mooted for the proposed health laws to cover most other industries. However in a last-minute concession to industry, when spokesmen began arguing that adherence to the 85dBa limit would be too expensive, the 1970 Occupational Safety and Health Act, OSHA[19] adopted the old 90dBa Walsh-Healey limit. As a result some disagreement resulting in controversy has arisen between the EPA's Office of Noise Control, and the National Institute for Occupational Safety and Health (NIOSH) which supports the Department of Labor's new act.

The above illustration may show that the devolution of expertise to specialist bodies who would advise the government on what to do only partially fulfils its intended purpose. The other more important non-technical decisions still have to be made. 'Expertise' is valuable, but it may in fact inhibit the ability of government to decide on their policy, because of the expert's failure to evaluate options from other more abstract perspectives. Even so, Commissions could well develop the proper policy-making skills at staff level. But they might still do no better than government departments in attracting suitable personnel.

Independence is a relative term. We ought not to forget that regulation continues to be dependent upon a procedure that tries to negotiate between affected parties and the rights of property holders. Additionally, the legislative impulse is generated from the administrators in charge of the agencies like the EPA in the USA and the HSE in Britain. In the long run there are few reasons to expect one group of administrators to make better social decisions than another group. Pressures of all sorts still impinge upon decision-makers.

11

PLANNING AGAINST NOISE

A clearly practical and logical solution to the problem of noise is to keep as far apart as possible those people or devices that create noise, and those people who are likely to hear the noise. This is an approach, of course, that the authorities, as far as they can, try to pursue under powers given to them by the central government. Planning measures are historically a better solution to solving problems arising from the social habitat than the prohibition of certain acts, and Parliamentary bills recognise this fact. For instance, the Road Traffic Regulations Act of 1967 and The Heavy Commercial Vehicles (Control and Regulation) Act 1973 give traffic authorities of county councils powers to control traffic (at least in non-trunk roads) in the interests of the general amenity.

But although the principle of spatial separation between noise-generating and noise-sensitive locale is simple enough, it has by no means always been applied in practice. This happens sometimes, for example through the failure to anticipate either the growth or the decline of traffic. It is a kind of problem that the planners have continually to contend with: noise must be abated, but where does the margin between ambient and obtrusive noise lie, and how will this margin shift in the future, and should the measuring yardstick differ between industrial and community locations?

What appears to be needed are an adequate predictive power that is precise enough to aid future planning policy and a rigorously formulated sound measurement system. As the NAC has pointed out, traffic management schemes are not always successful because current traffic flow prediction techniques are so inadequate.[1] This is amply illustrated by a report of a working committee issued to the Transport Minister in November 1977 which says that road traffic forecasts for the next thirty years are too high. It argues for a revised method of forecasting to include the impact of future oil shortages.[2]

Sizeable margins of health safety are more easily set by legislative bodies than by the courts. The problem with the courts is that they only respond, at least theoretically, to what is certain and can be proved, and not to what is speculative – it may be impossible to prove anything about noise and human wellbeing. The tendency not to adopt safety margins in the past has been reflected in the failure to take advantage of administrative planning potentials. It is a capitulation to the somewhat conservative and judicial principle that decrees that, until a hazard has been proved to exist, it cannot be controlled.

The principal aim of fighting noise pollution is containment and a gradual movement towards the reduction of levels. In most countries this entails control over land usage in addition to the implementation of building regulations and operational controls.[3] An early distinction made by planners between heavy and light industries was found to be unsatisfactory in practice because the noise range from a wide variety of premises was too broad to be properly classified. Hence planners arrived at the notion of *performance standards* in the zoning process, based upon a particular company's past record in the matter of noise emissions. The onus is placed upon the enterprise to predict its own noise levels, and to regulate itself. High emissions are tolerated as long as overall improvements are made.

Standards are also raised in accordance with the best, and most economic, means of control available. In Britain, following the recommendations of the NAC, Section 63 of the Control of Pollution Act 1974 permits the designation of certain Noise Abatement Zones in areas of mixed industrial and residential usage. The zone is brought about by the creation of a noise abatement order which will apply only to certain classes of establishments. The major change of attitude here is that it is no longer necessary to prove a nuisance before a local authority can require the implementation of remedial work.[4] The primary purpose of this power is the long-term control of noise from fixed, and primarily industrial, premises. The local authority is supposed to measure the noise coming from the designated area and to keep a public register of the recorded levels, which then act as optimal guidelines. The authority is also empowered to request a reduction in noise levels. In this way it is hoped that zoning will be effective in the long run.

Unfortunately there are one or two difficulties with both the theory and practice of zoning. The CBI complained about the Health Inspectorate's comparison between noise and smoke control practices. They pointed out that noise is not so readily identifiable and abated as smoke. In their view the success of 'smoke control areas' is due to the generous financial help given to householders for conversion to smokeless heating systems. The NAC agree with the CBI's comments, and refer also to the mobility of many noise sources. One can observe when smoke has been abated, but to check noise emissions requires very careful measurement.[5]

Moreover, the evidence about zoning in practice does not augur well. There is, for example, the failure to assess reasonable target levels that would have regard to the type and extent of the area, the proximity of dwellings, costs, etc.[6] Using additional powers, local authorities can refuse planning permission to an industrial development proposition that has paid scant regard to the matter of noise emissions. But, to the disappointment of the NAC, they frequently fail in this task.[7] It was partly for this reason that Part III of the 1974 Act strengthened the statutory features of zoning and similar measures.

Of over-riding importance to the planner is how well he can implement a *holding operation* (to contain present emission levels). Such an operation will entail expertise of the highest order. Local authorities must anticipate whether a factory *will* inflict excessive noise annoyance on nearby residents.[8] If the objectives of planning authorities are typically represented in the DOE's circular entitled 'Planning and Noise'[9] then one important skill is to prevent *new* commercial or industrial concerns from making any extra noise within a development scheme. It is important to envisage whether a factory will need to re-equip in the future so that the original process might be carried on more noisily, after which the authority can do nothing about it.

Local authority planners also need to cooperate with public health departments to ensure that any rise in ambient noise levels affecting residential areas are avoided. The problem for the embattled planner is that the introduction of new noise sources into an area is liable to result in a creeping growth in the overall level. For example, Table XIII shows that the introduction of a new source of 3dBa lower than the existing level would increase that level by 2dBa. The operative limits of the planner would be 75dBa by day and 65dBa by night.

Table XIII: Addition of sound levels (to the nearest ldBa). Source: Dept of Environment, 1973. Copyright, HMSO[10]

Difference in dBa between the two levels	Add to higher level
0	3
1	3
2	2
3	2
4	1
5	1
6	1
7	1
8	1
9	1
10	0

Planning depends to some extent on negotiative and consultative techniques. It is therefore rather unlike strict enforcement against which an individual can appeal but cannot effectively resist. In this sense it abides by our democratic principles, and the problems associated with it have some bearing on what was mentioned earlier.

Planning by consensus can be illustrated by drawing comparisons between the practices of communist and western countries. In the USSR buffer zones are instigated by the Sanitary Norms of 1956 and 1963 which require zones of various widths of those 'dirty' industries whose emissions include gases and particulates.[11] The precipitating motive for the creation of buffer zones will be the same for all countries, but the ethic behind the motive will vary. In Russia, entirely environmental or climatic factors, such as the direction of the prevailing wind, are operative. However, the planners in Britain adopt a modest stance. They admit that there is 'no absolute quantitative standard' with which they can judge whether any given noise is a 'nuisance'. The Standard BS4142, for example, aims to provide a means of assessing whether a given noise-level will give rise to complaints:

> The basic criterion of 50dBa is first corrected, if necessary, by the addition of 5 or 10dBa depending on the degree to which the particular factory fits into the character of the surrounding areas and whether people are used to this type of factory. A further correction is then made for the type of area itself, ranging from minus 5dBa for a rural area, to plus 20dBa for a predominantly industrial area with few dwellings. If the factory will operate only on weekdays between 8am and 6pm a further 5dBa is added, and if at night-times, 5dBa are subtracted.[12]

In other words, 5dBa is added to the 50dBa basic criterion for an established but untypical factory, and 10dBa is added for older established factories more in keeping with their surroundings. More 'corrections' are then made to take into account the 'noise climate'.

This method is aimed at finding which ambient noise levels would forestall complaints, but it also tries to assess whether the noise is typical of the circumstances. If the area is unusually noisy, complaints from the community may be expected if the corrected noise levels go beyond the criterion by 10dBa. However, the NAC admit there is a hazard in measuring noise in the presence of background emissions of a higher level. If this is worked out wrongly then an industrialist will rightly complain that his emissions are already below target level (but above the background level).

The GLC has used this method in its construction of a series of government-owned industrial plants. It has published guidelines showing how a hypothetical plant might be planned, as shown in Figure 9. Other methods of guessing the probability of complaints are occasionally derived from the benefit of hindsight. With each new highway built, additional useful experiences are gained about exactly how sensitive is a 'noise-sensitive development'.

But the consultative technique in planning goes further than this. The DOE planning circular of 1973 suggested that planning authorities should consult public health bodies about development applications where noise is a factor. The physician's role is thus seen to be as important as the technician's, since the information he receives from his patient can be incorporated into planning decisions. But this interactive approach to planning or zoning has its drawbacks. The planner cannot rely only upon the attitudes of residents. Industries, firms, bus companies, etc., also ought to be consulted. Unfortunately, should the consultative principle be spread wider, zoning may become vulnerable to the local influence of a large manufacturer.

In any event, considerable negotiative skills would be needed on the part of both the industrialist and the bureaucrat. The management of a company planning to build a plant has to search for empty buildings. When doing this it will have to consider the general requirement of each municipality in which the land or buildings are available.[13] The Council, for its part, must examine the possible disadvantages and financial costs to the area when it is presented with proposed building plans. The advantage of bringing in extra employment or facilities to the area must be considered. And the Council must also review its successes or failures in terms of containing noise from non-industrial sources. The business manager might well resent the suggestion that he should cooperate in reducing factory emissions when adjacent traffic appears to be quite unrestricted in noise terms. It is obvious that tolerance, skill and breadth of knowledge are most important to both sides. Plant management would need to have the fullest policy support from their superiors. They themselves would have to keep all levels of management informed about new arrangements that may be subject to more stringent abatement standards than before.

Overseas there are a variety of approaches to dealing with noise pollution, not all of course dissimilar to the British approach. Some countries simply set boundary limits, beyond which noise must not permeate. Such limits might take into account the contribution that ambient noise levels make to noise in general. Occasionally rigid restrictions imposed by central authorities are allowed to evolve into more flexible strategies within guidelines. There are a number of ingenious low-cost measures to combat environmental noise such as education campaigns, traffic management schemes, and product labelling schemes (where the manufacturer is required to state decibel levels on a product label).[14]

In Denmark, the town planner can reserve certain areas for industrial development.[15] This does not mean that all factories should be located together; rather, when application is made for a single factory, the total noise generated in the area is to be taken into account which might mean a relaxation of controls if traffic noise in the area is already high. Traffic noise levels from new roads are also permitted in accordance with a flexible emission criteria depending on whether the road passes near residential or industrial districts. The Environment Minister has powers of control over the design and manufacture of a wide range of domestic and industrial equipment.

In France, town planning law provides that buildings may be erected only in certain zones if they are indispensable to the existing population. The building licence for any property must indicate the position of the building in relation to noise sources. The town planning certificate must indicate that no noise problem exists, and that any requirements as to soundproofing must be met. Again, there is control over the design and building of equipment. Piston engine plant must not be used within 50m

(700ft) of houses and schools, etc., and should not emit sound over 85dB at a distance of 1m (3ft). Most other regulations to contain domestic and environmental noise are framed in statutes.[16]

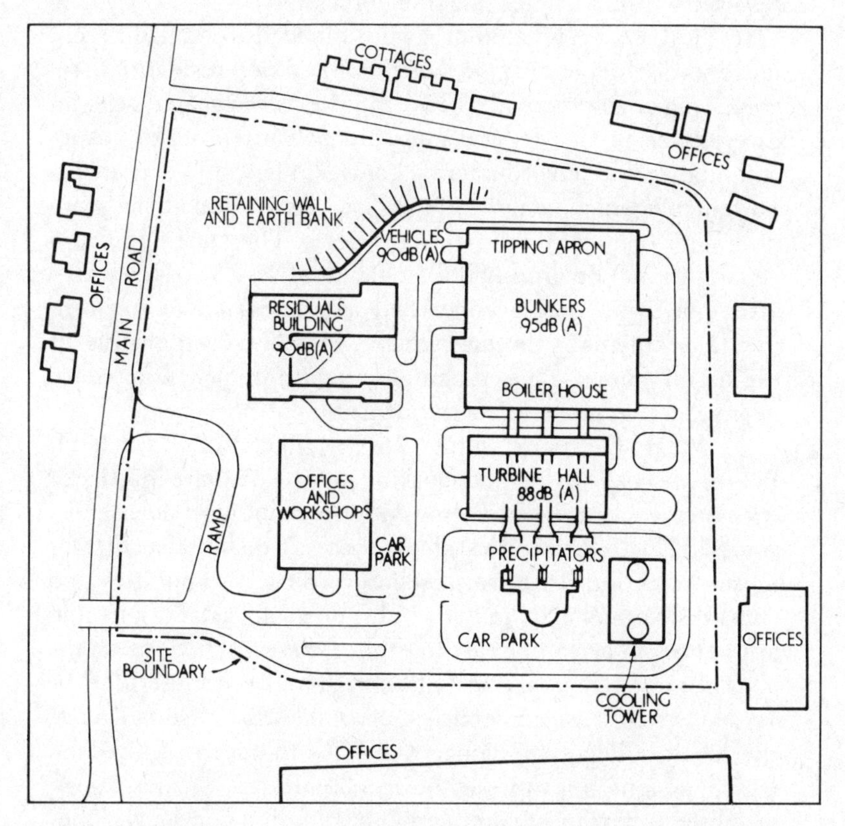

Figure 9 Plan of a refuse-treatment plant designed in accordance with Greater London Council noise abatement guidelines. Source: EPA (US), 1971

In the Netherlands, planning against noise may be taken by local authorities under the legal framework of the Physical Planning Act (WRO) (which makes no specific provisions).[17] A designation plan can regulate the use of land in line with the findings of multi-disciplinary researchers. Difficulties arise over territorial domains, as noise elements (i.e., from airports or motorways) may only partially be within the jurisdiction of a municipality or an entire region, which are attempted to be resolved within a framework of regional plans.

In the USA designated areas are allowed to emit different noise levels, depending on whether they are residential, recreational or commercial.[18] The 'top-down' approach is clearly in evidence in the Federal government's attempts to ensure compliance with environmental goals.[19] A few laws contain requirements which could lead to noise suppression at the planning stage of major public works projects. The most significant is the National Environmental Policy Act (NEPA), 1969, which attempts to provoke a concerned awareness about environmental problems by the government. This Act is responsible for the submission of 'environmental impact statements' in regard to any Act that might affect the environment.

In West Germany some *Lander* (like the Rhineland-Palatinate) have living and working areas which are interlaced with green belts roughly in line with the amount of noise emissions. In North Rhine-Westphalia there is a distancing edict for noise protection. In Bavaria self-governing communities and rural district councils may forbid industrial installations and appliances in certain areas. In other *Lander* restrictions on the operation of stationary installations may be imposed.[20] The West Germans use the concept of *Örtsublichkeit*—'suitability to the locale'. The same concept applies to the Swiss system, which assigns appropriate noise climates to various zones, weighted in favour of commercial or building concerns if they are felt to be desirable in an area.

12

NOISE AND THE LAW

The British legal system differs from those of other countries by the strong distinctions between remedies that may be gained through either the civil law of torts, or by Parliamentary law. The differences might be more clearly illustrated in the comparisons that can be made between common law nuisance, and statutory laws and statutory nuisance (where injurious processes are forbidden by law and cited in nuisance actions).

It was mentioned in Chapter 10 that the increasing trend towards change in environmental law was partly due to an awareness of certain inequities in implementing purely common law remedies. For instance, it has often been recognised that actions in common law tend to be brought about only by the rich or the poor, as the former can afford the high legal costs, and the latter qualify for legal aid. Most people must hence rely on public authorities to take action on their behalf by means of statutory law.

It should be noted, however, that 90% of confirmed cases of noise nuisance are dealt with outside the law. And about half of the complaints about noise were not confirmed by the local authorities as a statutory nuisance. This does not imply that most complaints are frivolous, but rather that the authorities are unable to pin the blame for the nuisance onto a particular source, or to satisfy the court that it constituted a public nuisance.

Many of the prohibitive measures against noise from most sources are now specified in the comprehensive 1974 Control of Pollution Act. This Act largely supersedes earlier legislation including the well known Noise Abatement Act of 1960, and many of the by-laws on the subject. The 1974 Act therefore covers most aspects of community noise nuisance, as well as construction site noise and the zoning arrangements of local authorities.

It is arguable that the legal trend since the 1960 Act is towards a reliance on statutory law, and that this tends to usurp the initiatory power of the common law remedies, where a person complains to a county court about a *public* or *private* nuisance. Nowadays a person can appeal instead to the local authority or the magistrates court. This is because under the 1974 Act it is officially recognised that unnecessary noise can be a nuisance, so the nuisance becomes *statutory* (whether it is public or private). However, the authority or the magistrate still has to determine whether a *prima facie* case for a nuisance exists before sanctions can be evoked against an offender. As the 1974 Act does not presume to define what a nuisance is, people have to fall back on the common law definition.

In law, noise can fulfil all the requirements of nuisance, which is a term that crops up in many statutes, in addition to the 1974 Control of Pollution Act. A distinction is made between a public nuisance (which is a crime) and a private one (which is a tort). There have been traditional problems in separating the dominion of the two, and the question is usually asked about how many individuals comprise 'the public' in order for an individual to be entitled to bring a public nuisance action. According to the judgment of Lord Denning, a public nuisance is one that is so widespread in its range and so indiscriminate in its effect that it 'would not be reasonable to expect one person to take proceedings on his own right'.[1]

But we should take care not to be misled by the word 'nuisance'. It is more than merely a bothersome irritation, since there must be evidence of damage that is not necessarily of a physical nature. Because English law is preoccupied with the defence of property rights it is not surprising that a nuisance will often be one where the plaintiff has been prevented from 'enjoying' his land.

Now, in the case of a plaintiff appealing to his local authority to abate the noise, we can see that they must themselves be satisfied that a nuisance exists as defined by the common law. They will do this by sending experts to investigate the matter, and if it is confirmed they will usually contact the person or persons responsible for the noise and attempt to get them to abate it voluntarily. Failing this, an abatement notice will be served on them requiring certain noise control works to be carried out; if they still fail to do this they may be prosecuted in a magistrates or High court.

A basic defence in a nuisance case is to try to show it was not a nuisance, or was a public rather than a private nuisance, or vice versa. Or the defendant might try to show that he still could not prevent the nuisance even with the exercise of 'all due care and attention'. Of course, if the noise had been caused by a third party trespassing on his land he again may plead that he was not liable for the nuisance. Occasionally, the defence is raised that the nuisance has existed continuously for twenty years or more and that the defendant has inherited a natural right to continue with it. In cases of statutory nuisance the defence could be used that 'the best practicable means' had been used to stem the noise, which in law is a defence distinct from pleading that due care and attention had been exercised.

It is no defence to say that the plaintiff came to the nuisance (as in the case of new housing being built near an existing factory) as business proprietors sometimes do. Neither can they plead that a particular firm is making or doing something useful for society, or that the nuisance only exists in concert with others.

Finally, we should note that in common law a noise-annoyed person has the right to take the law into his own hands by ridding himself of the noise without recourse to the courts. But such action is usually not recommended because it tends to remove the right of subsequent action in the courts.

In the case of road-traffic noise, the first point to be made is that the citizen does not have the right to bring nuisance actions in most cases concerning mobile sources of noise, and it is for this reason that state law is more clearly formulated and enforced.

The Motor Vehicles (Construction and Use) Regulations of 1973 is the main law which prohibits excessive vehicle noise, and it imposes stricter limits on the use of motor horns than before. The noise limits vary according to the type of vehicle and are set at a distance of not less than 17ft (5m) from the nearest point of the roadway being used at the time.

Motor Cycles	dB(a)
Under 50cc	80
50cc-125cc	85*
Any other motorcycle	89*
Cars	87
Light commercial vehicles	88
Heavy commercial vehicles	92

* 90dB(a) if first used before 1/11/70 [2]

Under a further law, the Road Traffic Act of 1974, it is now an offence to sound a motor horn at any time when a vehicle is stationary, except to warn another moving vehicle of an impending hazard. Horns must not be sounded by a moving vehicle in a restricted road between the hours of 11.30pm and 7am. Apart from the usual police, ambulance and other emergency vehicles, gongs, bells and sirens are not allowed. The previously tolerated multi-tone 'Alpine' horns have been forbidden since 1973.

Under an earlier Road Traffic Regulation Act of 1967, local traffic authorities have been given wide powers to control traffic in the smaller (non-trunk) roads in the interests of the comfort and convenience of the residential community. Further powers are vested in the Heavy Commercial Vehicles (Control and Regulations) Act 1973 to regulate the routing of heavy lorries. Also in 1973 the Noise Insulation Regulations were passed after the realities of six-lane motorways through dense outer suburban areas had become apparent, and when residents began to complain about the noise of new road building. The later Noise

Insulations Regulations additionally entitled house occupiers to compensation for the depreciation of property values arising from road and airport public building works, etc.

The control of aircraft noise is complicated by the fact that some airports come under the jurisdiction of local authorities and others under the British Airports Authority, which is partly a civil and partly a military organisation. However the main source of legislation is Section 29 of the Civil Aviation Act of 1971, which provides powers to restrict aircraft movements, and can specify flight paths for outgoing planes. During the months of April to October restrictions on night jet movements are in operation at Manchester and Luton airports, and throughout the year at Heathrow and Gatwick.

A number of other acts relate to noise in general. There is for instance the prohibition on the use of air powered tools that are not properly silenced under the Manchester Corporation (General Powers) Act of 1971. Of course, a great many anti-noise ordinances are contained in by-laws that are locally in force across the country under the enabling act of the Local Government Act, passed in 1933. The Home Office publishes model by-laws which are widely adopted by local councils. They cover the performance of music in public, loud radios, noisy animals and bird-scaring devices. However, most of these have been superseded by the 1974 Act, where new procedures are simpler and the penalties for failure have been strengthened.

Under Section 71 of the 1974 Act the Secretary of State has powers to issue codes of practice which are documents that explain how noise can be minimised. As in the case of occupational noise, which is discussed below, the government reserves the power to embody any such recommendations in follow-up legislation if necessary. Several codes have been, or are being, prepared to deal with noises emanating from, say burglar alarms, bird-scarers and noisy sports.

Finally, the 1974 Act bans the use of loudspeakers between 9pm and 8am, and during the day for advertising entertainments or for trade purposes. There is an exemption for merchants selling perishable foodstuffs.

We come now to one of the most important, and hitherto neglected, areas of noise nuisance; noise in factories. At the time of writing, late 1977, there is as yet no comprehensive law to limit the noise levels in factories and workshops on a national scale. There is a Code of Practice, issued by the Department of Employment (Code of Practice for reducing the exposure of employed persons to Noise, Department of Employment, HMSO, 1972), which recommends a general limit of 90dBa in industry. The Government has left itself the option of making the Code compulsory, and the committee responsible for the Code produced a further document in 1975 called 'Framing Noise Legislation' which discussed the feasibility of bringing in a new law based upon it.

The approach of the authorities to the problem has been one of gradual decision making and probing enquiry. Action in regard to occupational noise had to await the outcome of research carried out on behalf of the Industrial Injuries Advisory Council (IIAC). This research was undertaken jointly by the Medical Research Council (MRC) and the National Physical Laboratory (NPL) and the result of their findings was the well known Burns and Robinson report. It was focused on the question of eligibility for compensation under the terms laid down by the National Insurance (Industrial Injuries) Act of 1965. The Council asked whether there are degrees of hearing loss attributable to occupational noise which satisfy legal requirements for a loss of faculty of 1%.[3] In order to bring in a scheme of deafness compensation the council needed to find out if noise-induced hearing loss was a disease within the meaning of the act, and whether it was (also) a risk of occupation.

The common-sense view, as expressed by one witness to the Council's committee, which was accepted as a basic requirement for the award of disablement benefit by the Council, would be to regard hearing loss in terms of the hearing levels of patients attending clinics for the first time. In regard to the second question about occupational risks the Council concluded that deafness was one of the most likely diseases in Britain actually to be occupationally-induced, so in effect it more than fulfilled the conditions laid down in the Industrial Injuries Act. Hence, the question of whether hearing loss at work should be classed along with other industrial diseases was resolved, and awaited further governmental response.

In the absence of an overall law, there are only two statutory provisions which deal in specific terms with noise hazards as they affect workers. Regulation 44 of the Woodworking Machines Regulations of 1974 provides that, if a person is employed where woodworking machines are in operation, 'reasonably practicable' measures should be taken to reduce noise as much as possible where exposure was likely to reach 90dBa.[4] This law adds little to the common law of negligence but, like all of the 1974 environmental and occupational legislation, it introduces a criminal sanction. Where the noise level cannot be practicably reduced, and workers are affected, ear-protectors shall be provided and 'shall be used'. If they are not, for whatever reason, the occupier is in breach again.

The Agriculture (Tractor Cabs) Regulations of 1974 require tractors to be fitted with safety cabs which need a certificate of approval. The cab must not permit interior levels of more than 90dBa.

The earlier Factories Act of 1961 made no specific provisions in regard to noise, although the potential was always in existence: under section 76 of the Act, the Secretary of State may make such regulations where he feels that

> any manufacture, machinery, plant, equipment, appliance, process or description of manual labour is of such a nature as to cause risk of bodily injury to the person employed.[5]

However, this wording means that such regulations would be of a restricted nature, as it is unlikely that bodily harm would result from noise. It would thus have a limited application; it would, for example, only apply to a particular appliance or process. The courts have decided, in the case of Berry v Stone Manganese Ltd (1972), that the 1961 Act decrees that unwarranted noise emissions can make a factory an unsafe place to work in, and so becomes valid in statutory breach of duty cases.

Section 21 of the Offices, Shops and Railway Premises Act, 1963, confers powers to the Secretary of State to make regulations to protect persons from the risks of bodily harm or to health from noise, but no regulations under this act has been assumed. Noise levels with a serious risk are seldom found in premises to which the 1963 Act applies, but noise in some offices may be increasing with the introduction of computers and punching machines, etc.

The Health and Safety at Work Act 1974 is an important codifying act that embodies many of the provisions of earlier acts, although it makes no specific provision for measures against noise pollution. The general duties imposed on employers in Part II of the Act naturally encompass noise and vibration problems. It is the duty of the employer to ensure, as far as is reasonably practicable, the health, safety and welfare of workers at work, to consult with employees, and to make written statements of their safety policy arrangements.[6]

The above Act established two new organs, the Health and Safety Commission and the Executive, which are responsible for administering the provisions of the Act. In this regard, the aforementioned document 'Framing Noise Legislation', which was prepared by an Employment Department sub-committee *for* the Executive, would, if thought suitable, be recommended to become law by the Commission itself. Further regulations in terms of occupational noise are no doubt imminent in Britain, and will probably follow the suggestions made by the report. The report in fact recommended that four obligations be imposed upon employers, as follows:

1 that they carry out noise surveys in their places of work;
2 in the event of continuous noise levels exceeding 90dBa, all practicable means to reduce the noise must be taken;
3 failing this, excessive noise areas must be identified; and
4 all employees entering such areas must be provided with ear-protection.

If no new legislation is proposed, the Commission would like the Code to be made an Approved Code, so that failure to observe it would be a breach of Section 2 of the Health and Safety at Work Act.

In the USA, prior to 1970, the emphasis was on specific health and environmental problems that were the responsibility of independent agencies or Commissions. Each problem concerning noise tended to be considered in isolation from other health problems, and from the concerns of other states and of the Federal government. Yet the Federal government was unwilling to allow local regulations to curb noise arising from certain areas where 'interstate commerce' was affected. It felt that if the need arose to continuously take out court actions in order to nullify local noise laws, it would be more advisable to pass comprehensive, national laws. Furthermore, federal standards applied only to those corporations under contract to government agencies or which are federally supported.[7]

Later on, the administration's position hardened. Whilst reiterating the local nature of pollution problems, it suggested in 1971 that there were three areas (vehicles, construction equipment and internal combustion engines) that ought to be subject to 'Federal noise emission standards' under the coordinating aegis of the EPA.

During 1970, after the enacting of the National Environmental Protection Act (NEPA), Congress drafted and eventually enacted amendments to the Clean Air Act, Title IV of which was the Noise Pollution and Abatement Act of 1970. This act set up the office of Noise Abatement and Control (ONAC) within the EPA, and thus gave central focus to the Federal government's activities in noise abatement strategies. Even so, the ONAC was still primarily committed to carrying out investigations of various decibel levels in terms of community health. Its actual abatement authority tended to revert back to the affairs of federal department activities, rather than apply to industry as a whole.

In 1972, the Federal Noise Control bill was passed which gave the EPA coordinating responsibility for the many agencies involved in running the various programmes against noise-pollution. However, considerable autonomy is still allowed, and States and cities can pass their own laws. Most US cities have some sort of zoning scheme which regulates noise from industrial, commercial and domestic sources, some of which set very strict decibel limits. Other cities remove restrictions on industrial emissions where the ambient level is generally high. There is little consistency amongst municipalities in regulating noise from commercial establishments. In the past, noise control was left to the initiative of individual plaintiffs. In the judicial field, noise can be considered as a nuisance, physical trespass, 'inverse condemnation', and 'constitutional damage'. An important concept in US civil law is the 'gravity-utility' rule, where the balance of the harm caused to the plaintiff is weighed against the utility to the community of the processes or installations being sued against. The deficiency of this principle, from the plantiff's point of view, is that the courts often consider that his disamenity would not be as great as the harm caused to the investment entailed in existing businesses; in other words, the law is very much weighted in favour of the protection of economic rights. The application of environmental laws mainly to government contractors is also illustrated in US occupational noise law. In 1942 the Walsh-Healey Public Contracts Act was passed establishing minimum working conditions for employees of contractors supplying the federal government with materials and supplies. In May 1969, the Secretary of Labor gave the impetus to additional regulations concerning conditions in which contractors could hear conversations.

The scope of the occupational noise laws was considerably widened with the new Occupational Safety and Health Act of 1970, although the noise limits remained similar to the W H Act.[8] It authorised the Secretary of Labor to set mandatory occupational safety and health standards applicable to business 'affecting interstate commerce', and the broadness of its definitions meant that many more businesses were brought within government regulations. The remainder of industry, that did not come within the purview of the new Act, could be subject to civil or criminal sanctions. Nevertheless, under the Act a state could 'take over' the direction of a specific health matter such as occupational noise. Various other government departments can pursue either their own lead, or that of the Secretary of Labor, in regard to health laws. But the W H laws are the most widely accepted noise regulations within the federal programme.

Even so, the Walsh-Healey Act and the OSH Act differ in applicability and this has resulted in many more helpful noise abatement regulations under the newer Act. As the sanction of removal from a bidder's list is not available outside the framework of government contraction, the penalties under the 1970 Act use civil and criminal sanctions against those who infringe the law. Other agencies, such as the Bureau of Mines, and departments, use the Walsh-Healey standards as an abatement tool. Within individual states, limits are set 'pursuant' to national legislation, but similar to it. Seven other states have adopted regulations setting objective standards varying from 110 to 85dBa. Two states, Florida and New Mexico, have promulgated subjective standards. Almost every state has compensation schemes for occupational injuries and hearing loss.

In Holland, the authorities in the provinces and the local administrators look after the general interests within their borders, and after the particular interests of public utilities and industrial corporations. Certain matters are regulated separately, like laws which have the status of *lex specialis* in relation to the oldest environment act, the Nuisance Act. The Crown is empowered to promulgate general instruments of administration where Parliament has given the main outline. Policy is frequently prepared for promotion by individual ministers by policy planning boards such as the Council for Physical Planning.

The central authority has for some years been pursuing a policy of environmental hygiene and has been encouraging local authorities to do the same. The pollution problem is tackled either with fiscal or physical instruments, i.e., levies or mandates. Both are based on the 'polluter pays' principle which is intended to allocate financial responsibility and act as an incentive. Physical regulation is based on the 'best practicable means' approach, although the position is not yet reached where technical and scientific standards of permissability can be adopted as a basis for all regulations.

Planning measures for the control of noise may be taken under the Physical Planning Act (WRO). A designation plan regulates the use of land covered in the plan based on wide ranging research in order to formulate the best solutions for an area. Article 2 of the Nuisance Act forbids the establishment of certain installations capable of causing damage, danger or nuisance without a licence.

An amendment was recently passed to strengthen the Nuisance Act and to make provision for abatement of noise from construction equipment. Official concern has been expressed about the lack of responsible cooperation for environmental policy. In the meantime, in the absence of more unitary laws, recourse is often made to the civil law precedents such as the 1938 Unitas decision, or the 1966 Dordrecht court decision.

Under the Safety Act, an employer is required only to make available the means of personal protection, and an employee is obliged to make use of them. A recent amendment provides for supplementing the Act with effective measures against the propagation of noise. A more recent law prohibits the employment of young people in work areas where the pressure of sound in the auditory canal is above 90dBa without ear-protection.

In France, the creation of a separate ministry with special responsibilities for anti-pollution measures was set up in 1971. Earlier, the control of environmental nuisances was dealt with by different ministries like Agriculture, Public Health, Interior, etc. A later decree of June 1974 created the Ministry of the Quality of Life, which included nature, sport, leisure and tourist concerns. There is, however, no codification of the scattered texts on environmental protection matters because perhaps of the fear of legislative rigidity. A special code dealing with noise has recently been published.

Under a Town Planning Law, the building licence for any property must show the position of a building in relation to noise sources, and prove that soundproofing regulations have been met. For plant built since 1969 these should be 85dBa, except for communal circulation areas for which it is 70dBa. Building-site machinery is subject to specific Decree 69–380 of April 1969 in regard to the soundproofing of site machinery.[9] There is a complex system of controls for construction plant which differ according to whether they were built before or after May 1973. Prior to the May 1973 date equipment can be used only at a distance of 50m from residential buildings. Since May 1974 silencers must have been fitted for the intake and exhaust stroke of industrial motors.

Decree no 69–348 of April 1969 requires a firm to maintain noise at levels compatible with the maintenance of workers' health by the usual sound reducing techniques. Advisory Circular TE34/71 of November 1971 sets out noise limits and gives recommendations on audiometry. The decree allows a factory inspector to require that a noise survey be made by an approved organisation at the firm's expense.

In Germany, there is a power-sharing system between the State and the *Lander* for environmental policy formation and enactment. At neither level is there total concentration of all powers relevant to the environment in one Ministry, and problems of coordination arise amongst ministries. The federal authority enjoys overall planning status, and devises frameworks for Land legislators to elaborate upon.[10]

The Environmental Programme of the Federal Government of September 1971 imposes objectives and development standards for the country. The Federal Emission Control Law of 1974 codifies a large number of disparate laws giving protection against noise. Orders to curb noise emissions may be made under this law, or under appropriate laws for certain categories of premises under the various *Lander*.

Federal legislation to protect workers from industrial noise is currently in draft form and is imminent. The North Rhine-Westphalia is the only *Lander* presently to require surveys on the lines recommended in the *Richtlinie* (an advisory code on audiometry and hearing protection measures) to be made at noisy factories. A special data sheet must be completed and copies sent to a central processing centre. In addition, specified medical data forms must be completed and sent to a central office, and workers must be informed of noise hazards involved and of the necessity of hearing tests and personal protection.

In Denmark, the Ministry of the Environment has 3 out of 8 offices involved in anti-pollution matters, and can lay down subordinate legislation. It can also intervene in the enactments of subordinate bodies like Directorates. In spite of the standardisation of general provisions, there is some variation from one municipality to another.[11] Guidance is given to planning authorities in a booklet, 'External Noises from Industry', which sets levels not to be exceeded in six types of area, ranging from rural to urban-industrial. General planning legislation offers a wide range of possibilities for controlling the siting of a plant. Chapter 5 of the Environmental Protection Act is of importance when deciding upon plant sitings, and factors such as the importance to the community of local industries are taken into account. Where industrial noise is transmitted to surrounding dwellings it should not appear to exceed 30dBa with the windows shut. Decisions about siting permissions are influenced by the number of existing industries in the area, and by whether background traffic is higher than the proposed installation.

No rules exist over the construction of plant and equipment, but in Guidance 3/1974 on external noises from factories, issued by the Directorate, authorities giving approval should lay down limits. Controls can if needed be introduced by the Minister of the Environment under article 6 of the Environmental Protection Act.

Occupational noise provisions are covered in the Occupational Safety, Health and Welfare Act no 226 of June 1954, and later amended, Section 19 of which specifies injurious limits. Similar provisions exist for commercial, clerical and agricultural workers. Binding requirements of the Danish Inspection Authority (*Arbejdstilsynet*) cover the building and installation of machinery, construction and work. The Inspection Authority also sets limits, subject to constant review, which are currently an equivalent continuous sound level of 90dBa determined in accordance with ISO 1999. It also requires 'reasonable' measures to be taken by employers, failing which protective equipment is to be supplied which employees are obliged to wear. Audiometry is not mandatory.

EPILOGUE

So what of the future of our noise polluted world? If one were thinking about some of the excellent research that is currently being conducted in our universities and institutes then there is much hope for a quieter future. But noise, we must remember, is not just an acoustic problem, but a societal one. We have seen in Part I that there is a distinct correlation between trends towards increasing urbanisation and personal mobility and the level and pervasiveness of noise. So regardless of long-run engineering advances it remains clear that there is some inevitability about the amount of noise in society. For example, although there is some success in reducing the noise of individual types of aircraft, the number of aircraft *movements* increases at 7% a year which tends to more than counterbalance the effect of emission reductions.

Indeed, there remains the danger that some disingenuous officials will try, by mathematical manipulation, to make the situation seem better than it is. The variety of noise indices and the nature of their formulations make this a distinct possibility. The PNdB index might well show a gradual improvement as noisier aircraft are phased out, although measurements like the NNI, which reflect both noise and the frequency of flights, might prove the opposite case.

And we must look carefully at the distinct advantages of not interfering too much with the noisy status quo. Britain is still a relatively prosperous country. Although the worst sorts of noise arise from aircraft and road traffic, a great deal of the indices of economic advancement depends upon the profitability of the mobility industry. The package-tour industry is pleased to predict continued growth; in the late 1960s, the number of bookings for arranged holidays abroad rose by 20% a year, and was still rising by 11% in 1975. Naturally, the authorities are placed in the universal dilemma about how to satisfactorily work out the trade-off between industry (and economic growth and jobs) and environmental concerns (and the residential victims of airport expansion). Our study of the administrative function has shown that the authorities have evolved specialist agencies to try to successfully tackle this dilemma.

Inevitably, manufacturers are worried about their competitive vulnerability, and they will look askance at the idea of raising prices (of cars) or fares (of aircraft flights) to cover the cost of fitting extra silencers. And although plastic components with their better damping qualities, ought to be more widely used in car manufacture, there is much investment in metal working plant. In fact, in the interests of keeping costs low, there is a trend towards lightness which means that parts of motor engines are made thinner so that they have lower mass and provide less sound insulation.

But how is the designer and inventor coping with the noise problem? What of the future of engineering modifications to contain noise? The noisier turbojets are already being phased out, but largely because they use more fuel than the high-bypass turbofan aircraft. By the 1980s only about 60% of them will remain in service in Britain. Quieter aircraft such as the Lockheed Tristar, the DC10 and the latest version of the Boeing 747 have already entered service. An NAC subcommittee has suggested that in the not too distant future aircraft noise will not be intrusive beyond a short distance outside of airport boundaries. This, of course, remains to be seen.

In the meantime scientists continue to find ways of quietening the noisier low-bypass ratio engine. There are noise certification requirements now in force for aircraft of new designs, which the older designs will also have to abide by. Rolls Royce are confident that existing retrofitting techniques will enable them to reduce the noise of the RB211 by a further 5dB, and they are receiving government support. They propose to modify the air intake, to install more sound absorbent linings, and to change the design of the turbine and tailpipe, all of which should gain another substantial reduction of 5dB. The Concorde manufacturers have also spent about £40 million to investigate quieter operational techniques and modifications to engines and airframes.

Of course, innovative and even revolutionary changes have been mooted by many brilliant and forward-thinking people in the fields of acoustics and public planning. In order to actually implement some of them, vast sums of money would need to be spent. In the abstract they look promising: one example is the 1972 suggestion by the Noise Abatement Society that all existing airports be closed and replaced with offshore airports, the first of which could be Foulness, off the Essex coast. There would need to be a great number of such airports, so that prospective air passengers are situated no more than an hour away from the nearest, and then could be fed to them by a system of, say, advanced passenger trains.

When the present moratorium on motorway building is lifted in the hopefully more prosperous future, it has been suggested, all new long distance roads should be cut into deep valleys with sound deflecting concrete screeds overhanging the tops of the valley walls. The GLC has recently suggested that some 680km (425miles) of special routes for heavy lorries to and from ports should be built, to which lorries over 16 tons and 11m (36ft) in length could be restricted.

Dramatic new discoveries in acoustics have recently drawn the attention of the world to the concept of 'anti-sound'. Professor J. E. Ffowcs Williams of Cambridge University believes it is possible to create a sound pressure wave which would be identical to an unwanted sound, but exactly out of step, its troughs coinciding with peaks in the sound wave so that the two waves would cancel each other out. In theory, a device could be constructed that could be fitted to heavy lorries, or installed inside houses situated next to motorways so that noise in both cases could be literally blotted out. A certain amount of success has already been achieved with regard to low frequency noise, but it is more difficult to get a precisely synchronised 'anti-sound' to relate to all frequencies in the spectrum.

Finally, the universities have never been short of creative talent, and no doubt it is only a matter of time before some of their experimental successes are put to good practical use for the benefit of society. They always have been in the past. What is clearly needed is more cooperation between the innovative scientist and the more utility-conscious designer of consumer appliances and vehicles. Enforcement by the authorities might in fact be a necessary evil. If silencing devices are officially required, the consumer will be obliged to pay the higher price resulting from the extra research and the fitting of them.

Unfortunately quietude, like so many of the good things in life, has its price.

REFERENCES

1 The Growth of Noise Pollution

1 *The Times*, 16 November 1977
2 OECD, 'Motor Vehicle Noise', Paris, 1971
3 Pearce, David, 'Social Cost of Noise', OECD, Paris, 1976, p4
4 Bugliarello *et al.*, *The Impact of Noise Pollution*, Pergamon Press, 1976, p387
5 *Ibid*, p50
6 'Noise', US Department of Labor Guide to OSHA Standards, 1972, p3
7 Hines, V. A., *Noise Control in Industry*, Business Publications, 1966, p133
8 Taylor, Rupert, *Noise*, Pelican Books, 1975, p86
9 *Noise*, *op. cit.*, p3
10 'Noise in Public Places', Noise Advisory Council, HMSO, 1974, p11
11 *Ibid*, p12
12 Aldous, Tony, *Battle for the Environment*, Fontana, 1972, p231
13 'Noise in the Next Ten Years', NAC, HMSO, 1974, p*v*
14 'Bothered by Noise?', HMSO, NAC, 1975
15 Blair, Thomas L., *The International Urban Crisis*, Hart-Davis, MacGibbon Ltd, 1974, p106
16 *New Society*, 27 May 1971
17 'Review of Aircraft Routing Policy', NAC, HMSO, 1974, p12
18 Morgan, Elaine, *Falling Apart*, Souvenir Press, 1976, p173
19 'Noise in Public Places', *op. cit.*, p16
20 *Ibid*, p16
21 *London Evening News*, 20 December 1977
22 'Noise in Public Places', *op. cit.*, p14

2 The Price of Mobility

1 Aldous, Tony, *Battle for the Environment*, Fontana, 1972, p67
2 Rivers, Patrick, *The Restless Generation*, Davis-Poynter, 1972, p44

3 'Noise in the Next Ten Years', NAC, HMSO, 1974, p*v*
4 Blair, Thomas L, *The International Urban Crisis*, Hart-Davis, MacGibbon (Granada), 1974, p99
5 Bugliarello *et al.*, *The Impact of Noise Pollution*, Pergamon Press, 1976, p86
6 *New Society*, 2 September 1971
7 NASA, 'Transportation Noise Pollution, Control and Abatement', Langley Research Center, 1970
8 Blair, *op. cit.*, p76
9 OECD, 'Urban Traffic Noise', Paris, 1970
10 Brook, Peter F., *Problems of the Environment*, Harrap, 1974 p142
11 NAC, *op. cit.*, p4
12 Mishan, E. J., *Growth – the Price we Pay*, Staples Press, 1969, p68
13 Allaby, Michael, *Inventing Tomorrow*, Abacus books, 1977, p160
14 *New Society*, 2 September 1971
15 Brooks, *op. cit.*, p143
16 Wilson, 'Committee on the Problem of Noise', HMSO, 1963
17 Rivers, *op. cit.*, p137
18 Bugliarello *et al.*, *op. cit.*, p379
19 Rivers, *op. cit.*, p138
20 *Ibid*, p141
21 NAC, *op. cit.*, p8
22 *New Society*, 8 August 1968
23 Kryter, K. D., *The Effects of Noise on Man*, Academic Press, 1970, p126
24 Brooks, *op. cit.*, p146
25 *New Society*, 10 October 1974

3 Industrial Noise

1 'Neighbourhood Noise', NAC, HMSO, 1971, p17
2 *Ibid*, p27
3 Osborn, W. C., in *Environmental Pollution Control*, edited by McKnight, Marstrand & Sinclair, Allen & Unwin, 1974, p277
4 Davies, D. L. National Physical Laboratory (NPL), 'The Control of Noise', HMSO, 1962, p8
5 NY Daily World report, 22 May 1975
6 HM Factory Inspectorate, 1974 annual report, HMSO, p72
7 Kerse, L. A., *Noise and the Law*, Oyez Publications, 1974, p110
8 Osborn, W. C., *op. cit.*, p278
9 Wakstein, Charles in *Environmental Pollution Control*, *op. cit.*, p233
10 Sims, L. C., in *Aspects of Environmental Protection*, edited by Jenkins, S. H., IP Environmental Ltd, 1973
11 Burns, William, *Noise and Man*, John Murray, 1973

4 The Costs of Noise

1 Rivers, Patrick, *The Restless Generation*, Davis-Poynter, 1972, p137
2 *Ibid*, pp138–9
3 Mishan, E. J., *Growth – the Price we Pay*, Staples Press, 1969, p38
4 Pearce, David, 'The Social Cost of Noise', OECD paper, 1975, p11
5 Bugliarello *et al.*, 'The Impact of Noise Pollution', Pergamon Press, 1976, p151
6 *Ibid*, p155
7 Rivers, *op. cit.*, pp137, 138
8 *St Louis Globe* report of 25 August 1972, quoted in NVB October 1972
9 Lipscombe, in *Aspects of Environmental Protection*, edited by S. H. Jenkins, IP Environmental Ltd, 1973
10 Cited by Pearce, *op. cit.*, p28
11 EPA document, 'Noise from Industrial Plants', EPA/NTID 300.2, p246
12 *Ibid*, p214
13 Wakstein, Charles, 'The Noise Problem in the USA', 5th International Congress for Noise Abatement, London, 1968
14 Wakstein, Charles, in *Environmental Pollution Control*, edited by McKnight *et al.*, Allen & Unwin, 1974
15 OECD, 'Economic Implications of Pollution Control', 1974, p30
16 *Ibid*, p20
17 Personal communication
18 EPA document, 'The Economic Impact of Noise', EPA/NTID 300.14
19 Wakstein, *op. cit.*, p251
20 Report in NVB April 1975
21 Private communication
22 EPA, Federal Register, Vol. 39. No. 244, P4 3807
23 Bolt, Beranek and Newman report 3246 to EPA, 1974
24 Pearce, *op. cit.*

5 Noise as an Energy Form

1 'Industrial Noise', Wolfson Unit, Institute of Sound & Vibration Research, 1976, pp1–2
2 Henry, T. A., 1976 Brunel lecture, British Association for Advancement of Science
3 Archer, H., in *Aspects of Environmental Protection*, edited Jenkins, S. H., IP Environmental Ltd., 1973, p298
4 D. W. Robinson, 'The Control of Noise', HMSO, 1962, p8
5 *Ibid*, p11

6 Henry, *op. cit.*, p8
7 Rupert Taylor, *Noise*, Pelican books, 1975, p99
8 Richards, E. J., ISVR University of Southampton internal document, 'Mechanical Noise Sources', p22-2
9 'Aircraft Engine Noise Research', NAC, HMSO, 1974, p2
10 Richards, E. J., *op. cit.*, p22-1
11 Middleton, A. H., ISVR University of Southampton internal document, 'Current State of the Art – future prospects,' 1975

6 Noise, Hearing and Well-being

1 'Industrial Noise', Wolfson Unit, Institute of Sound & Vibration Research, University of Southampton, 1976, p1.5
2 Botsford, J. H., *Journal of the Acoustics Society of America*, vol 42, 1967, p810
3 The British Code of Practice for reducing workplace noise, HMSO, 1972
4 Report by the Industrial Injuries Advisory Council (IIAC), 'Occupational Deafness', HMSO, 1973, p10
5 Noise & Vibration bulletin report, April 1975
6 EPA document, Public Health and Welfare Criteria for Noise, 550/9–73–002, 1973, pp3–5
7 *Ibid.*
8 'Noise in the Next Ten Years', 1974, NAC, HMSO, p2
9 'Should the NNI be Revised?', NAC, HMSO, 1972, p1
10 Bugliarello *et al.*, *The Impact of Noise Pollution*, Pergamon, 1976, p403
11 Burns and Robinson, 'Hearing and Noise in Industry', HMSO, 1970, p5
12 Middleton, A. H., ISVR University of Southampton internal document, 'Current State of the Art – future prospects', 1975, p18-1
13 Department of Employment, 'Code of Practice for reducing the exposure of employed persons to noise', HMSO, 1975
14 'Noise and the Worker', HMSO, 1968, p7
15 Glorig, A., Summerfield, A., and Nixon, J., paper in *Proceedings of the 13th International Congress on Occupational Health*, NY, p150
16 IIAC, *op. cit.*, p3
17 Burns and Robinson, *op. cit.*

7 Noise and Disablement

1 Quoted by Charles Wakstein in *The Public and National Noise Standards*, edited by Golovin, Brookings Institute and Harvard University, 1968, p228

2 *Ibid*, p227
3 Miller & Jacoby, *JASA*, Volume 27, 1958, p338
4 Atherley & Purnell in *Industrial Safety Handbook*, edited by Handley, McGraw-Hill, 1969, p332
5 Trittipoe, W. J., *Journal of the Acoustics Society of America*, *31*, p244
6 Federal Register, EPA, volume 39, No 244, p43803
7 EPA document 550/9, 'Public health & welfare criteria for noise', 1973, pp2–17
8 *Ibid*
9 Report of the IIAC, 'Occupational Deafness', HMSO, 1973, p14
10 Valcic, I., ILO paper, 1974, p66
11 'A Survey of regulations to prevent noise risks', ILO paper, Geneva, 1974
12 Duerdon, Cyril, *Noise Abatement*, Butterworth, 1970, p76
13 Davies, D. L., National Physical Laboratory (NPL), 'The Control of Noise', HMSO, 1962, p320
14 Kryter, K. D., *Journal of the Acoustics Society of America*, 1966
15 Carpenter, A., NPL, *op. cit.*, p304
16 *Ibid*, p305
17 *Ibid*, 303
18 *Ibid*, p303
19 *Ibid*, p302
20 Committee on the Problem of Noise, 'Final Report', HMSO, 1963
21 Teichner, Arces & Reilly, *Ergonomics 6*, 83, 1963
22 'Notes on Occupational Deafness', HMSO, 1974

8 Designing Out Noise

1 Berry & Horton, *Urban Environmental Management*, Prentice-Hall, 1974, p289
2 Hines, V. A., *Noise Control in Industry*, Business Publications, 1966, p40
3 Lockwood, D., ILO paper, 1974, p43
4 Taylor, Rupert, *Noise*, Pelican, 1975, p132
5 Henry, T. A., British Association of Science paper, Brunel University, 1976, p8
6 Hines, *op. cit.*, p72
7 Bugliarello *et al.*, *The Impact of Noise Pollution*, Pergamon Press, 1976, p147
8 *Ibid*, p102
9 Rivers, Patrick, *The Restless Generation*, Davis-Poynter, 1972, p148

9 Controlling Industrial Noise

1 HM Factory Inspectorate, 1974 annual report, HMSO, p73
2 *Ibid*

3 'Noise and the Worker', HMSO, 1968
4 'Industrial Noise – the Conduct of the Reasonable and Prudent Employer', Wolfson Unit, University of Southampton, 1976, p5–3
5 EPA document, 'An assessment of noise concern in other countries', EPA/NTID 300.6, 1971, p246
6 Bell, Alan, 'Noise – An occupational hazard', WHO, 1966, p64
7 Hines, V. A., *Noise Control in Industry*, Business Publications, 1966, p81
8 Atlas Copco promotional literature, 1975
9 EPA document, 'Noise from Industrial Plants', EPA/NTID 300.2 p246
10 Bolt, Beranek and Newman, Report 3246, 'Economic Impact Analysis of Proposed Noise Control Regulations', 1975, prepared for OSHA
11 'Noise and the Worker', *op. cit.*, p9
12 EPA 300.2, *op. cit.*
13 *Ibid*
14 Martin, A. M., 'Occupational Hearing Loss and Hearing Conservation', ISVR paper, University of Southampton, pp23–4
15 *Ibid*, pp23–7

10 Noise and the Administrative Function

1 Wakstein, C., *Environmental Pollution Control*, McKnight, Marstrand and Sinclair, 1974, p210
2 Bell, Alan, *Noise*, WHO, 1966, p103
3 Bugliarello *et al.*, *The Impact of Noise Pollution*, Pergamon Press, 1976, p330
4 Harvey and Hallett, *Environment and Society*, Macmillan, 1977, p112
5 Colliard, Albert, 'The Law and Practice Relating to Pollution Control in France', Graham Trotman Ltd, Commission of the European Community, 1976, p12
6 Meyer, A. F., *Journal of the Acoustics Society of America*, March 1972, *3*, pt 1, pp800–2
7 Vickers, Sir Geoffrey, in *Style in Administration*, edited by R. A. Chapman, Allen & Unwin, 1971, p101
8 Harvey and Hallett, *op cit.*, p87
9 Savage, Burke *et al.*, 'Economics of Environmental Protection', Houghton and Mifflin, 1974, p2
10 Osborn, W. C., in *Environmental Pollution Control*, *op cit.*, p275
11 'Neighbourhood Noise', NAC, HMSO, p31
12 Self, Peter, *Administrative Theories and Politics*, Allen and Unwin, 1973, p56

13 Lord, William B., in 'Economics and Decision Making', Food and . resource Economic Department papers, University of Florida, 1971, edited by Connor and Loehman, p124
14 Hufschmidt, M. M., in *Economic Analysis of Environmental Problems*, (ed) E. S. Mills, National Bureau of Economical Research, 1975, Columbia University Press, p438
15 Meyers, S., *op. cit.*, p45
16 Duerdon, Cyril, *Noise Abatement*, Butterworths, 1970, p61
17 *Ibid.*
18 Annual report of the Scientific Advisor, GLC, Scientific Branch, 1973
19 Osborn, *op. cit.*, p278

11 Planning against Noise

1 'Noise in the Next Ten Years', NAC, HMSO, 1974, p5
2 *Sunday Times*, 13 November 1977
3 Jones, D. K., MSc paper, International Congress on Acoustics, Geneva, 1977
4 Large, J. B. and Taylor, Rupert, 'Inter-Noise', ISVR, University of Southampton, 1975, p716
5 'Neighbourhood Noise', NAC, HMSO, 1971, p35
6 Archer, H., in *Aspects of Environmental Protection* (ed) S. H. Jenkins, IP Environmental Ltd, 1973, p305
7 'Neighbourhood Noise', *op. cit.*, p21
8 *Ibid*
9 'Planning and Noise', DOE, circular 10/73, 1973, p4
10 *Ibid*
11 EPA report, 'An assessment of Noise Concern in other Nations', EPA/NTID 300.6, 1971, p237
12 *Ibid*, p238
13 EPA report, 'Noise from Industrial Plants', EPA/NTID 300.2, 1974, p209
14 Jones, D. K., *op. cit.*
15 Jensen, C. H., 'Law and Practice Relating to Pollution Control in Denmark', Trotman Ltd, Committee of the European Community, 1976, p158
16 Colliard, Albert, 'The Law and Practice Relating to Pollution Control in France', Graham Trotman Ltd, Commission of the European Community, 1976
17 McLoughlin, J., 'Law and Practice Relating to Pollution Control in EEC Countries', Trotman, *op. cit.*
18 Bugliarello *et al., The Impact of Noise Pollution*, Pergamon Press, 1976, p344

19 Osborn, W. C., in *Environmental Pollution Control* (ed) McKnight *et al.*, Allen and Unwin, 1974, p283
20 McLoughlin, *op. cit.*, p322

12 Noise and the Law
1 See 'Handbook of Noise & Vibration Control' (ed) R. H. Waring, Trade & Technical Press, 1974
2 'Bothered by Noise', NAC Pamphlet, 1975, p2
3 Report of the IIAC, 'Occupational Deafness', HMSO, 1973, p14
4 'Industrial Noise – the Conduct of the Reasonable & Prudent Employer', Wolfson Unit, ISVR, University of Southampton, 1976, pp6–11
5 Kerse, L. A., *Noise & the Law*, Oyez Publications, 1974, p111
6 *Ibid*, p116
7 EPA report, 'Laws & Regulations for Noise Abatement', NTID 300.4, pp1–33
8 *Ibid*, pp1–34
9 Colliard, A., 'Law and Practice relating to pollution control in France', Commission for the European Community, Graham Trotman Ltd, 1976, p124
10 Steiger & Kimminich, 'Law and Practice Relating to Pollution Control in the Federal Republic of Germany', CEC, *Ibid*, p11
11 Jensen, C. H., 'Law & Practice Relating to Pollution Control in Denmark', *Ibid*, p3

Select Bibliography

Anthrop, Donald F. *Noise Pollution* (Lexington Books, 1973)
Bell, Alan. *Noise* (World Health Organisation, 1966)
Bugliarello *et al. The Impact of Noise Pollution* (Pergamon Press, 1976)
Burns and Robinson, 'Hearing and Noise in Industry' (HMSO, 1970)
Burns, William. *Noise and Man* (John Murray, 1973)
Cunniff, P. F. *Environmental Noise Pollution* (John Wiley & Sons, 1977)
Duerdon, Cyril. *Noise Abatement* (Butterworth, 1970)
Jenkins, S. H. (ed). *Aspects of Environmental Protection* (I.P. Environmental Ltd, 1973)
McKnight, Marstrand and Sinclair (ed). *Environmental Pollution Control* (Allen & Unwin, 1974)
Taylor, Rupert. *Noise* (Pelican Books, 1975)

INDEX